ADVANCES IN BIOCHEMICAL ENGINEERING

Volume 7

Editors: T. K. Ghose, A. Fiechter, N. Blakebrough

Managing Editor: A. Fiechter

With 112 Figures

Springer-Verlag
Berlin Heidelberg New York 1977

ISBN 3-540-08397-9 Springer-Verlag Berlin Heidelberg New York
ISBN 0-387-08397-9 Springer-Verlag New York Heidelberg Berlin

Library of Congress Catalog Card Number 72-152360
Printed in Germany

Typesetting, printing, and bookbinding: Brühlsche Universitätsdruckerei Gießen.
2152/3140-543210

Contents

Bubble Column Bioreactors

Tower Bioreactors Without Mechanical Agitation

K. Schügerl, J. Lücke
Institut für Technische Chemie der Technischen Universität Hannover,
D-3000 Hannover

U. Oels
Bayer AG, Werk Uerdingen, D-4150 Krefeld-Uerdingen

Contents

Summary

The present article investigates the behavior of bubble column bioreactors with yeast culture media in the absence of cells. To aid in the assessment of these reactors the following properties were estimated and partly theoretically treated: relative mean gas hold-up, bubble swarm velocity, bubble size, gas/liquid interfacial area, energy requirement for aeration, oxygen transfer coefficient across the gas/liquid interface and backmixing in the liquid phase. All of these properties are strongly influenced by the composition of the culture medium and the type of aerator. It is shown that in bubble column bioreactors, in the absence of antifoam agents and with low viscosity culture medium, high oxygen transfer rates can be achieved at low energy requirement. By application of multistage columns particular properties of the bubble column can be varied significantly. A comparison of bubble column reactors with mechanically agitated, as well as with air-lift bioreactors, indicates that bubble columns are economical reactors, especially for aerobic cultivations.

Introduction

Bubble column reactors are popular in the chemical industry because of their versatile use and economical advantages i.e., low investment costs due to their simple construction and low variable costs of production due to low energy requirement of their operation, which is maintained by fluid dynamical mixing and dispersion of the phases. All these advantages are also valid for their application in biotechnology. However, with few exceptions, bubble column bioreactors have not been introduced into industry yet because of the lack of the necessary know-how in their design and operation.

A further reason for the delay of the application of bubble column bioreactors is an economic one. The replacement of sterile stirred tank bioreactors by new, more economical, reactor types is often unattractive due to high initial costs and long durability of the existing equipment.

In stirred tank bioreactors the construction of the equipment, especially that of the aerator, is not as decisive as in the case of bubble column bioreactors, since the performance of the stirred tank bioreactor can be improved up to a limit by an increase of the mechanical energy input. The lack of such a safety factor increases the risk of the applied bubble column bioreactor not giving the required performance. Therefore more careful and accurate design than for stirred tank bioreactors is necessary to lower the risk. The necessary data for this design exist only partly. The aim of this paper is to present more basic data for bubble column bioreactor design and operation, especially with regard to SCP production.

1. Application of Bubble Column Bioreactors in Industry

Large scale industrial application of bubble column bioreactors is rare [1, 2]. After a series of patents had been granted [8–13] on so-called 'tower fermentors' for the production of alcoholic liquids, in particular beer, the first application was realized [3, 4, 14–17], in pilot plant and then in production-scale.

The commercial production of beer in tower fermentors has been carried out in continuous operation over a period of many years. The beer produced is for all practical purposes indistinguishable from that produced in the conventional batch fermentation. For the successful continuous operation of a tower bioreactor it is essential to use a flocculent yeast, which is easy to separate from the beer at the tower head, since otherwise the yeast would be washed out and an insufficient yeast concentration maintained. The mean (wet) yeast concentration 25% w/w is generally attained with values as high as 30–35% w/w at the bottom and as low as 5–10% w/w at the top. These bioreactors are also applied in the U.K. in vinegar production [4]. Newer applications of such bioreactors without mechanical agitation have been worked out for SCP-production. Since the planned commercial units are larger than the usual bioreactors, by the application of standard stirred tank bioreactors, one would run into several difficulties due to the necessity for heat removal in external cooling loops with sufficiently high pumping rates, and to the intensive aeration as well as agitation required. Their high energy requirement would prevent the economical operation of a large commercial plant. To avoid these difficulties, especially the external cooling of the fermentation medium which is necessary to remove the great amount of heat produced by mechanical stirring, new pneumatic bioreactors were developed. Air-lift bioreactors became popular. One of the first patents for the use of this bioreactor was granted to Lefrancois in 1955 [18], and several units which are in use are described in the literature [19–25]. ICI has developed a bubble column bioreactor with external recycling [the pressure cycle fermentor (PCF)] which has operated very satisfactorily in pilotscale production of ca. 1000 tons per year protein [26]. Pilot plant air-lift bioreactors have been applied by BP Proteins Ltd. [37]. The Kanegafuchi Chemical Industry Co., Ltd., Japan, has also developed a bubble column bioreactor with external recycling for n-paraffin solutes [28]. The air-lift bioreactor of Gulf Research and Development Co., Pittsburgh, is supplied by a draft tube and applied in the semicommercial petroprotein unit at Vasco, California [29]. A bench-scale bubble column bioreactor has been used for yeast production in Prague [5].

Different modifications of bubble column bioreactors, e.g. multistage tower reactors with a mechanical stirrer, have also been developed [6, 7].

To further extend the application of these pneumatic tower bioreactors (bubble column reactors with and without recycling) more data are needed which are evaluated under fermentation conditions.

2. Properties of Bubble Columns and Their Characterization

The main task of bubble columns for aerobic cultivation is the dispersion of air in the liquid to maintain the high oxygen transfer rates, OTR's, necessary to high productivity. The OTR depends on the overall volumetric mass transfer coefficient and the oxygen driving force. Since in common cultivation systems the gas side mass transfer resistance can usually be neglected, one considers only the resistance in the liquid phase. This assumption is made in the present paper.

The following definitions are used:

The oxygen driving force is the difference between the oxygen concentration in the liquid at the interface C_L^* and in the bulk C_L. It is assumed that C_L^* can be calculated by the partial pressure of oxygen in the gas phase and the Henry coefficient H_e of oxygen, i.e. at the interface distribution equilibrium of oxygen prevails.

Instead of the over all mass transfer coefficient the individual mass transfer coefficient of the liquid phase k_L is used, which is defined by Eq. (1):

$$Q_{O_2} = k_L a' (C_L^* - C_L) \tag{1}$$

where $Q_{O_2} = \left(\frac{\text{mass } O_2 \text{ transferred}}{\text{time}}\right)$

C_L, C_L^* = concentration of O_2 in the liquid bulk phase and/or at the interface $\left(\frac{\text{mass } O_2 \text{ dissolved}}{\text{liquid volume}}\right)$

a' = gas/liquid interfacial area

$$k_L = \frac{Q_{O_2}}{a'(C_L^* - C_L)} = \left(\frac{\text{volume of liquid}}{\text{interfacial area} \times \text{time}}\right) \text{ or}$$

$$k_L a = \frac{k_L a'}{V_L} = \frac{Q_{O_2}}{V_L} \frac{1}{C_L^* - C_L} = \left(\frac{1}{\text{time}}\right)$$

with $a = \frac{a'}{V_L}$ specific interfacial area $\left(\frac{\text{gas/liquid interfacial area}}{\text{volume of liquid}}\right)$.

The correlations for the mass transfer coefficient are usually expressed in dimensionless form:

$$Sh = C_1 (SC)^n (Gr)^m \tag{2}$$

where $Sh = \frac{k_L \cdot d_B}{D_L}$ Sherwood number

$Sc = \frac{\nu_L}{D_L}$ Schmidt number

$Gr = \frac{d_B^3 \rho L \Delta\rho g}{\eta_L^2}$ Grashoff number

d_B = mean bubble diameter

D_L = diffusivity of O_2 in the liquid phase

$\nu_L = \frac{\eta_L}{\rho_L}$ kinematic viscosity of the liquid phase

η_L = dynamic viscosity of the liquid phase

ρ_L = density of the liquid phase

$\Delta\rho = \rho_L - \rho_g$

ρ_g = density of the gas phase

g = acceleration due to gravity

n, m and C = constants.

Because bubbles of different size behave dissimilarly the experimental correlations are valid only for a given range of d_B: e.g. for small bubbles ($d_B < 2.5$ mm) Calderbank *et al.* [30] gave Eq. (2) with

$$C_1 = 0.31, \qquad n = 0.33, \qquad m = 0.33$$

and for large bubbles ($d_B > 2.5$ mm) with

$$C_1 = 0.42, \qquad n = 0.5, \qquad m = 0.33.$$

According to these equations k_L is independent of bubble size within these two ranges and depends only on the physical properties of the system [31].

The specific interfacial area "a" can be calculated for a swarm of nearly spherical bubbles by Eq. (3):

$$A = \frac{a'}{V} = \frac{6\,E_G}{d_s} = a \cdot (1 - E_G). \tag{3}$$

In Eq. (3) E_G is the mean relative gas hold up defined by Eq. (4):

$$E_G = \frac{V - V_L}{V} = \frac{H - H_L}{H} \tag{4}$$

where V = volume of the bubbling layer
V_L = volume of the bubble free liquid layer
H = height of the bubbling layer
H_L = height of the bubble free liquid layer
a = a'/V_L specific interfacial area

d_s is the "Sauter" mean or surface volume mean diameter:

$$d_s = \frac{\sum_1^N n_i d_i^3}{\sum_1^N n_i d_i^2} \tag{5}$$

where n_i = the frequency of the bubbles with the diameter d_i.

According to Oels [32] a simple relation between E_G and d_s holds:

$$E_G = C_2 Fr^p \tag{6}$$

where Fr = $\frac{w_{SG}}{\sqrt{g d_s}}$ Froude number
w_{SG} = superficial gas velocity
C_2 and p are constants.

Therefore the specific interfacial area A depends only on w_{SG} and d_s:

$$A = \frac{C_3 {w_{SG}}^p}{g^{0.5p} {d_s}^{1+0.5p}} \tag{7}$$

where C_3 is a constant.

According to Eq. (7) A can be enlarged either by increasing w_{SG} or by diminishing d_s. Economical production demands as low energy input as possible to keep low the amount of heat produced by energy dissipation and the cooling capacity needed to remove this heat.

The increase of w_{SG} means higher compression energy (and sometimes heavier foam problems), the decrease of d_s can be achieved in different ways. Since the bubble diameter plays a decisive role in aerobic fermentors, especially in bubble columns with regard to the OTR, its dependence on the most important design and process parameters has to be considered.

In stirred tank reactors the air is injected into the liquid at the base of the mixing vessel by a simple gas intake system. The dispersion of the injected gas phase is achieved by

mechanical agitator which produces dynamic pressure by means of turbulence.
In the turbulent field the large bubbles are unstable, they disintegrate. Their size is controlled by the dynamical equilibrium bubble size. In bubble column reactors often flat gas distributors (perforated or porous plates) are applied which produce small bubbles. The initial size of these bubbles depends on the forces acting on them during their formation. Several investigations have been carried out on single bubble formation at orifices and nozzles [44]. Since perforated plate distributors are multi-orifice and porous plate distributors are multinozzle systems, the initial bubble size can be calculated by the relations developed for single bubbles, as long as the interaction of the bubbles is low and the bubble formation frequency remains below a limit. At high bubble formation frequency coalescence can occur during the formation (pairing of bubbles [45–48]). However, if the oxygen requirement is low, relative low gas flow rates are used, therefore it is possible to apply the well known relations developed for single bubbles [44] to calculate the initial bubble size.
In general, bubble formation occurs in two stages, i.e. the expansion and ascending stages. During the expansion stage the bubble is kept on the nozzle- or orifice-opening. It grows by the inflowing gas. According to the static theory [44], the first stage is finished when the buoyancy force becomes greater than the forces which act downwards on the bubble. At this point the bubble begins to ascend. During the ascending stage the bubble remains connected with the nozzle and/or orifice opening by a tail, which has the diameter of the opening. Across this tail the bubble is fed further by gas. The bubble formation is stopped by the disconnection of the tail. According to the dynamical theory [49] the lift-off of the bubbles from the orifice (or nozzle) is caused by the inward radial motion of the liquid which narrows the tail of the bubble up to detachment.
For constant gas flow rate Q, which is applied in the present investigation, the final volume of a single bubble V_B is given by Eq. (8):

$$V_B = V_E + Q\,t_c. \tag{8}$$

Here t_c is the duration of the ascending stage and V_E the volume of the bubble at the point of lift-off [44]:

$$V_E^{5/3} = 0.047\,\frac{Q^2}{g} + 2.41\,\frac{\nu_L}{g}\,Q\,V_E^{1/3} + 3.14\,\frac{D_{\ddot{o}}\sigma}{g\rho_L}\,V_E^{2/3}. \tag{8a}$$

The final volume of the bubble can be estimated by Eq. (8b):

$$r_E = \frac{B}{2\,Q(A+1)}\,(V_B^2 - V_E^2) - \frac{C}{AQ}\,(V_B - V_E) - \frac{3\,G}{2\,Q\,(A-\frac{1}{3})}\,(V_B^{2/3} - V_E^{2/3}) \tag{8b}$$

where $r_E = \left(\frac{3}{4\,\pi}\right)^{1/3} V_E^{1/3}$

$A = 30.6\, r_E\, \nu_L/Q$

$B = 1.45\,\frac{g}{Q}$

$C = \frac{4.55\,D_{\ddot{o}}\sigma}{\rho_L Q}$

$G = 3.51\,\nu_L.$

Equation (8) can only be applied in the range in which separate bubbles are formed. With increasing gas flow rate a transition at the aerator from bubbling gas into gas jet occurs. The critical flow rate of this transition is given by Eq. (9a) and/or (9b):

$$Q'_{cr} = 20.4 \frac{\sigma}{\rho_L} \sqrt{\frac{D_ö}{g}} \quad (cm^3/s) \tag{9a}$$

according to Brauer [108] and

$$Q_{cr} = \sqrt{\frac{2\,\pi^2 D_ö^3\,\sigma}{16\,\rho_G}} \quad (cm^3/s) \tag{9b}$$

according to Ruff [109]. Here ρ_G = the density of the gas phase.
These two equations yield rather different results. However, one can consider Q'_{cr} as the upper and Q_{cr} as the lower limit of the critical flow rate. If a gas jet forms in the laminar liquid at the orifice or nozzle it disintegrates in a given distance from the opening due to the instability of the gas-liquid interface. For the systems investigated in the present paper the viscosity terms can be neglected, hence the inequality (10a) prevails [110]:

$$\frac{\alpha\,\rho_L}{\eta_L} \gg k^2, \tag{10a}$$

where α = growth rate of disturbance cm^{-1}
k = wave number of disturbance cm^{-1}.

The general stability equation then simplifies to [110]:

$$\alpha^2 = \frac{\sigma(1 - k^2 a^{*2}) k a^*}{\rho_L a^{*3} K_0(ka^*)/K_1(ka^*)}, \tag{10b}$$

where a^* = jet radius,
K_0 = modified Bessel function of the second kind or zeroth order,
K_1 = modified Bessel function of the second kind of first order.

This equation was first derived by Rayleigh [111]. The controlling wave length corresponds to the dimensionless wave number $(ka^*)_{max} = 0.485$.
Equation (10b) can be applied, if

$$\frac{\sigma \rho_L D_ö}{\eta_L} > 36. \tag{10c}$$

With the media, perforated an porous plate used in this investigation inequality (10c) is always fulfilled.
Since the size of the bubbles formed from laminar cylindrical gas jets is controlled by the amplification of disturbances which result from surface instability, one can calculate the bubble volume, if one assumes that it is equal to the volume of the cylinder having the radius of the jet and the length λ, the acutal wave length of dominant wave [112]:

$$V_B = \pi\, a^{*2} \lambda \tag{10d}$$

where $\lambda = \frac{2\,\pi a^*}{ka^*}$,

hence

$$V_B = \frac{2\,\pi^2 a^{*3}}{(ka^*)_{max}}\,. \tag{10e}$$

By substituting the nozzle or orifice radius for the jet radius one obtains bubble volumes which agree fairly well with the experiments. Relation (10c) can be applied up to the gas flow rates where at the aerator no turbulence prevails.

In the presence of local turbulence at the aerator the ratio of the dynamic pressure force of the local turbulence to the interfacial force controls the bubble size. Therefore the initial bubble size at the gas distributor is controlled by the buoyancy and interfacial forces [Eq. (8)] at low gas flow rates (bubbling gas range), by the instability of the gas/liquid interface of the gas jet [given by Eq. (10b)] at intermediate gas flow rates, and by the ratio of the dynamical pressure force of the local turbulence to the interfacial force at high gas flow rates.

However, this initial bubble size is not necessarily preserved in the entire column. The ascending bubbles coalesce, if the initial bubble size is smaller than the local dynamical equilibrium bubble size in the column and the coalescence it not supressed, alternatively the bubbles disintegrate if the initial bubble size is larger than the local dynamical equilibrium bubble size. In systems with hindered coalescence the bubbles formed at the gas distributor can be preserved, therefore the size of the bubbles can be smaller than the dynamical equilibrium size, if the initial bubble diameter is below the dynamical equilibrium diameter in the column.

In pure liquids the coalescence/redispersion rate is high, therefore the bubbles quickly attain the equilibrium size. In this case and in systems with gas distributors which produce initial bubble sizes larger than the equilibrium size, the bubble diameter is not influenced by the gas distributor plate. The bubble size is controlled only by the dispersion equilibrium in the column. The dispersion equilibrium is reached when the ratio of dynamic to surface tension forces has a particular value which is characteristic for the system. This force ratio is given by the Weber-number *We*:

$$We = \frac{\tau d_B}{\sigma} \tag{11a}$$

where τ = dynamic pressure,
σ = surface tension.

For dynamic equilibrium Eq. (11b) holds:

$$We_{eq} = \frac{\tau d_{B\,max}}{\sigma} \tag{11b}$$

where $d_{B\,max}$ the maximum possible diameter of the bubble which can survive at dynamic equilibrium in a flow or turbulent field of dynamic pressure τ. In a one stage bubble column, in which the gas bubbles ascend due to the buoyancy forces with the relative velocity w_R with respect to the liquid the *We*-number is given by [38]:

$$We^2 = \frac{\rho_L w_R^2\, d_B}{2\,\sigma} \tag{12}$$

for nearly spherical bubbles with a diameter of d_B, if one can neglect the viscous and inertia forces. Equation (12) holds for low viscosity liquids, as investigated in this paper, and for systems in which the bubble movement is not influenced by external forces (except gravity).

The Bond number, *Bd*, accounts for the gravitational and surface tension forces:

$$Bd^2 = \frac{\rho_L g d_B^2}{4\,\sigma}. \tag{13}$$

Berghmans [38] evaluated the boundary between the stable and unstable regions for bubble swarms as function of the *We*- and *Bd*-numbers by neglecting the viscous and intertia forces.

Calculating *We*- and *Bd*-numbers by means of the measured ρ_L, w_R, σ, and d_B the position of the bubbles on the stability diagram can be estimated [50]. If they are in the stable region the dynamical coalescence-redispersion equilibrium is not important. The bubble diameter is not influenced by $d_{B\max}$, but by the initial bubble diameter at the gas-distributor. If the bubbles are at the stability boundary, dynamical equilibrium prevails and $d_B \cong d_{B\max}$.

It is difficult to estimate the dynamic pressure τ of Eq. (11) for a complex turbulent flow. Such turbulence prevails, e.g. in a bubble column with nozzle aeration, near to the nozzle. Turbulent flow produces primary eddies which have a scale of similar magnitude to the dimensions of the main stream.

These large primary eddies are unstable and disintegrate into smaller eddies until all their energy is dissipated by viscous flow. During the transfer of the energy from primary eddies to small eddies the directional nature of primary eddies is gradually lost [40]. According to Kolmogoroff [39] the smallest eddies which are responsible for the energy dissipation are statistically independent of the primary eddies and have locally isotropic character. The scale of these smallest eddies l is given by Eq. (14):

$$l = \frac{\eta_L^{3/4}}{\rho_L^{1/2}} \left(\frac{E}{V_L}\right)^{-1/4} \tag{14}$$

where $\frac{E}{V_L}$ is the rate of energy dissipation per unit volume of the liquid.

If one assumes that in the presence of a bubble the local structure of the turbulence does not alter, the maximum stable diameter of the bubble is given by the ratio of the attacking shear stresses and the surface tension resisting the deformation of the bubble, i.e. by the *We*-number:

$$We = \frac{\overline{u^2(d_{B\max})}\,\rho_L\, d_{B\max}}{\sigma} \tag{15}$$

where $\overline{u^2(d_{B\max})} = \overline{(u_1 - u_2)^2}$

and u_1 and u_2 are the local velocities of the liquid at the maximum distance of $d_{B\max}$.

If $L \gg d_B \gg l$, where L the scale of primary eddies and l the scale of the smallest eddies $\overline{u^2(d_{B\max})}$ can be calculated by Eq. (16) [41]:

$$u^2(d_{B\max}) = C_4 \left(\frac{E}{V_L}\right)^{2/3} \left(\frac{d_{B\max}}{\rho_L}\right)^{2/3}. \tag{16}$$

Putting Eq. (16) into Eq. (15) and comparing it with Eq. (11) one obtains the theoretical relation (17) for τ

$$\tau = C_4 \rho_L \left(\frac{E\, d_{B\max}}{V_L \rho_L}\right)^{2/3} \tag{17}$$

and for d_{max}

$$d_{max} = C_5 \frac{\sigma^{0.6}}{\left(\frac{E}{V_L}\right)^{0.4} \rho_L^{0.2}} \tag{18}$$

where C_4 and C_5 are constants.
In a given system (C_5, ρ_L and σ are constant) $d_{B\max}$ depends only on the rate of energy dissipation $\frac{E}{V_L}$, i.e. on the power input per unit volume of the liquid. Therefore one can produce small bubbles ($d_{B\max}$ is small) by a high rate of energy dissipation. However, high $\frac{E}{V_L}$ means also high power input per unit volume, i.e. high energy requirement and high variable costs. Economical operation demands low power input.
If the coalescence rate in the fermentation medium is low, it is more economical to apply the energy in a small volume at the site of bubble formation, i.e. to apply a high rate of energy dissipation locally and to retain overall a relatively low energy requirement [42]. Thus one can form small bubbles due to small $d_{B\max}$ in this volume. By delayed coalescence this small bubble size can be preserved and a high specific interfacial area can be achieved with relatively low energy input. Coalescence can be delayed or completely supressed by additives (e.g. C_1–C_2 alcohols etc.) which are often used as substrates. The influence of the substrate on "a" should be considered, both when the substrate is selected and on the specification of "a".
Longitudinal mixing also influences the operation of continuous bubble column fermentors.
In continuous stirred tank fermentors it is assumed that the mixing is "perfect", therefore the ideal continuous stirred tank reactors (CSTR) model can be applied. According to this model the agitation is sufficient to assume homogeneous conditions so that the composition of the effluent from the vessel is always the same as the composition of the contents. The material balance over the vessel with respect to the cell mass X in the vessel is given by Eq. (19)

$$\frac{dX}{dt} = D\,(X_0 - X) + \mu X \tag{19}$$

where X_0 = cell concentration in the feed,
D = $\frac{E}{V_L} = \bar{t}^{-1}$ dilution rate or reciprocal residence time
F = feed
V_L = volume of the liquid in the vessel.
Under steady state conditions:

$\frac{dX}{dt} = 0$ and with

sterile feed, $X_0 = 0$, the specific growth rate μ equals the dilution rate:

$$D = \mu. \tag{20}$$

If the intensity of the mixing is less than "perfect", the actual wash out rate of organism is less than μ unless the liquid culture is (frequently or continously) reinoculated. This question was treated by Erickson *et al.* [33, 34]. On the other hand the utilization of the substrate is much better, if the intensity of the longitudinal mixing is low. Thus the optimal operation of a reactor and its productivity depends on the longitudinal mixing of the phases in the reactor.

The intensity of axial mixing is usually described by longitudinal dispersion models [36, 37] or back flow cell models [35, 36]. If the oxygen partial pressure in the gas phase changes only slightly, one can neglect the longitudinal dispersion in the gas phase and consider it only in the liquid phase. For this case the estimation of the longitudinal fluid dispersion and the application of a one-phase model is sufficient to characterise the axial mixing in the fermentor. In the present paper only the axial mixing in the liquid phase is considered and only the longitudinal dispersion model is applied.

The dispersion model is described by the following dimensionless equation, derived from an unsteady state material balance on the tracer component.

$$\frac{\partial C^*}{\partial \theta} + \frac{\partial C^*}{\partial x^*} = \frac{1}{Pe} \frac{\partial^2 C^*}{\partial x^{2*}} \tag{21}$$

where C^* = dimensionless concentration
θ = dimensionless time
x^* = dimensionless axial distance.

Equation (21) is used to estimate the model parameter [37, 43].

This model is based on the assumption that the two transport mechanisms bulk flow and longitudinal dispersion are independent of the position in the reactor. The dimensionless model parameter, the Peclet number *Pe*:

$$Pe = \frac{w_L L}{D_{L\,\mathrm{eff}}}$$

indicates the degree of mixing within the reactor. Here w_L is the effective flow rate of the liquid, $D_{L\,\mathrm{eff}}$ the effective longitudinal diffusivity and L is the test section.

In view of the foregoing considerations the following parameters are going to be used to characterise the bubble column fermentor:

E_G average relative gas hold up
w_R relative velocity of the bubble swarm
d_s "Sauter" mean diameter of bubbles
We Weber number of the bubble column
Bd Bond number of the bubble column
a, A specific interfacial area gas/liquid
$k_L a$ volumetric mass transfer coefficient of oxygen across the gas/liquid interface.
k_L mass transfer coefficient of oxygen across the gas/liquid interface
Sh Sherwood number

$\frac{E}{V_L}$ or $\frac{E}{V}$ rate of energy dissipation necessary to produce the specific interfacial area a.

D_L coefficient of longitudinal dispersion and/or back mixing in the liquid phase.

Several investigations have been carried out for the estimation of E_G, A, $k_L a$ and D_L in bubble columns of pure liquids (mostly water). Most recent evaluations for E_G [51], A [52] and Sh [53] are given by Gestrich and for D_L by Eissa [54] and Todt [36]. However, fermentation media have a complex composition and no general correlations are available for liquid mixtures.

3. Systems and Procedures

a) Biological System

Since the general aim of the current investigations is the optimization of yeast (*Candida boidinii*) production from alcohol in bubble column fermentors the culture medium given below was used with various substrates.

1 g	KH_2PO_4/l
2 g	$KHPO_4$/l
2 g	$(NH_4)_2SO_4$/l
2 g	$(NH_4)_2NO_3$/l
1 g	$Na_2HPO_4 \cdot 2H_2O$/l
0.2 g	KCl/l
0.2 g	$MgSO_4 \cdot 7H_2O$/l
0.5 mg	H_3BO_3/l
0.04 mg	$CuSO_4 \cdot 5H_2O$/l
0.1 mg	KI/l
0.2 mg	$FeCl_3 \cdot 6H_2O$/l
0.4 mg	$MnSO_4 \cdot H_2O$/l
0.4 mg	$ZnSO_4 \cdot 7H_2O$/l
0.2 mg	Ammoniumheptamolybdat/l.

The yeast (*Candida boidinii*) and the composition of the above culture medium originate from "Gesellschaft für Biotechnologische Forschung e. V." Stöckheim [55], our partner in cooperative research.

The following substrates were used in conjunction with the medium:

methanol
ethanol and/or
glucose
and for comparison
n-propanol
n-butanol and/or
10% Na_2SO_4.

The latter corresponds to the commonly used sulphite oxidation system for the estimation of the specific gas/liquid interfacial area [56].

To investigate the influence of the phosphates and the different substrates separately the following media were used:

1. demineralized water as reference liquid,
2. H_2O/NOP (culture medium without phosphates),
3. H_2O/NMP (culture medium with phosphates),
4. H_2O/CH_3OH (0.5–20%)
5. H_2O/C_2H_5OH (0.5–15%)
6. H_2O/n-C_3H_7OH (0.5–15%)
7. H_2O/n-C_4H_9OH (0.5–20%)
8. H_2O/glucose (1.0–20%)

and the combination of 2. and/or 3. with the substrates 4. to 8.
The methanol and ethanol substrates were also combined with cells (at different cell densities). Solutions 1., 2., and 3. without and with methanol and ethanol substrates were used at pH = 2 and pH = 7.
Some properties of these solutions are given in Table 1. The pH-value of the fermentation medium was varied between 2–7, but no influence was found on the most important properties of the bubble column.

Table 1. Properties of some model media at 25 °C

Medium	Density (g/cm^3)	Surface tension (dyn/cm)	Viscosity (cS)	fluid number $- \cdot 10^{-10}$
H_2O	0.997	70.0	0.889	5.33
0.5% CH_3OH	0.997	64.1	0.913	3.88
1% CH_3OH	0.998	62.9	0.930	3.42
2% CH_3OH	0.996	62.6	0.960	2.99
0.5% C_2H_5OH	0.999	58.6	0.914	2.95
1% C_2H_5OH	0.998	57.2	0.936	2.50
2% C_2H_5OH	0.996	53.8	0.981	1.73
H_2O/Salt	1.005	71.5	~1	3.67
2% CH_3OH/Salt	1.0	65.7	0.98	3.07
1% glucose	1.002	73.0	0.908	5.34
2% glucose	1.006	73.8	0.932	5.37

cS = centistokes

$$\text{fluid number} = \frac{\rho_L \cdot \sigma^3}{g \cdot \eta^4}$$

b) Apparatus

Figure 1 shows a schematic view of the apparatus, column B with height of about 4 m and diameter of 14 cm was used to absorb the O_2 in the liquid medium and column C (with the same diameter and height) to desorb it by N_2. Hence the liquid enters at the bottom of column B oxygen free and the oxygen concentration in the liquid medium increases along the column.
The longitudinal O_2 concentration profiles in the liquid medium were measured by 10 oxygen electrodes. To depress the disturbances due to the entering liquid flow on the bubble flow, a stainless steel screen (I in Fig. 1) was inserted to give a flat radial O_2

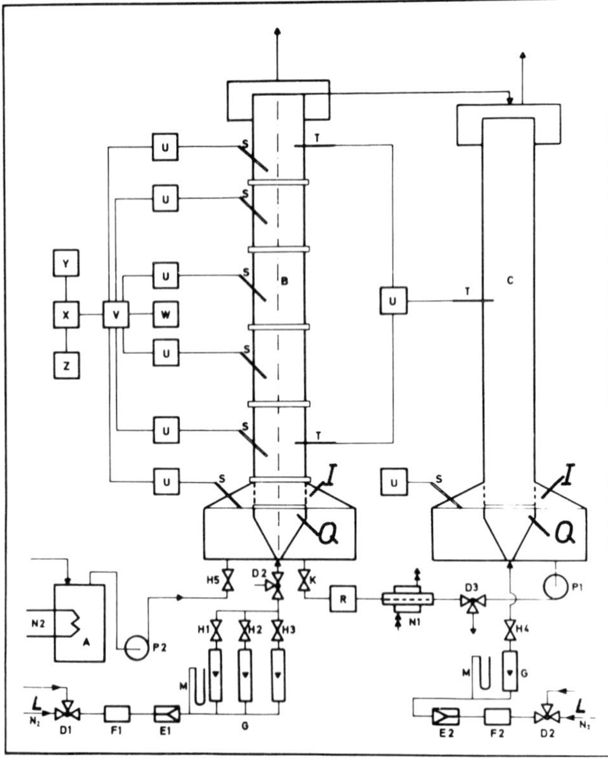

Fig. 1. Schematic diagram of apparatus.
A liquid tank
B O_2-absorption column
C O_2-desorption column
D three-way-magnetic valve
E pressure regulator
F filter
G flow meter
H valve
I stainless steel screen (size of mesh 200–800 μm)
K magnetic valve
L air in
M pressure gauge
N heat exchanger
N_2 nitrogen in
P pump
Q gas distributor plate
R inductive flow meter
S O_2-electrode
T resistance thermometer
U amplifier
V Selector of measuring channels
W timer
X analogue-digital converter
Y addressing unit
Z tape punching device

concentration profile at a distance of 15 cm from the gas distributor. The longitudinal concentration profile in the column up to this point was used for the estimation of the volumetric mass transfer coefficient.

Four different types of gas distributors were used: a Cr-Ni stainless steel porous plate with a thickness of 2 mm and a mean pore diameter of 17.5 μm (manufactured by Sintermetallwerke Krebsöge) and a perforated stainless steel plate of 1 mm thickness with 180 bore holes of 0.5 mm diameter and free surface area of 0.23%, and two two-component nozzles: a stainless steel ejector nozzle with nozzle diameter of 3 mm and an injector nozzle of 4 mm diameter.

The ejector nozzle was developed by R. Sinn und O. Nagel, BASF, Ludwigshafen [42, 98]. It consists of a central cylindrical nozzle for the liquid phase with a diameter of 3 mm and a coaxial annular nozzle with a slit width of 1.2 mm for the gas phase. A liquid free jet with a nozzle outlet velocity of 5–100 m/s is formed, which expands into the so called "momentum exchange tube" (with inside diameter of 26 mm and length of 180 mm). The expanding gas forms large bubbles periodically which are dispersed in the momentum exchange tube in a plane normal to the jet axis where the liquid bounces against the tube wall and a steep shear velocity is formed [98].

The injector nozzle was developed by M. Zlokarnik, Bayer AG, Leverkusen [99]. It is actually a water jet pump. The water jet, which formed by a small nozzle of 4 mm diameter enters into a slightly larger converging-diverging nozzle and entrains the gas being around the jet. The gas sucked into the converging-diverging nozzle is dispersed there [98].

We thank BASF, Ludwigshafen and/or Bayer AG, Leverkusen, for placing the ejector and/or injector nozzle at our disposal.

Oilfree air and purified nitrogen were used. The liquid medium was transported by a glass centrifugal pump (P1 in Fig. 1) (GPB 3/V/25T of Quickfit Glastechnik) between the columns B and C in a closed circuit. The rate of liquid flow was measured by an inductive flow meter (R) (Krohne, Duisburg). Teflon membranes 25 μm and/or 50 μm and 0.1 n KCl electrolyte solution were used in the O_2 electrodes (WTW, Beckmann and GMBF, Stöckheim). The analogue signals of the 10 electrodes were converted by a data logger and stored on tape.

The rates of liquid flow and/or gas flow were varied between 410 l/h and 1,200 l/h and/or 0.1 Nm^3/h and 3 Nm^3/h. The temperature was held constant (25 °C). Additional experimental details are given in reference [57].

c) Measuring Methods

The gas hold up was estimated by the simultaneous interruption of the gas and liquid flows by two magnetic valves, the cell concentration was measured by a photometer (Zeiss RMQ 2), and the alcohol concentration in the fermentation medium was analysed frequently by gas chromatography (Shimadzu, Poropak Q 150 °C). The fermentation medium was renewed every day. The radial O_2 concentration profiles were measured at the same time as the longitudinal O_2 concentration profiles to confirm the validity of the onedimensional dispersion model.

The axes of the assembled O_2-electrodes had an angle of 60° to the axis of the bubble column and all 10 electrodes could be moved up to 30 cm along their axis. Each of the electrodes were put in three positions. By that 30 points of the longitudinal concentration profiles could be measured. The concentration of oxygen at each point was recorded for 5 min.

The relative velocity of a bubble swarm w_R is given by Eq. (22):

$$w_R = \frac{w_{SG}}{E_G} - \frac{w_{SL}}{1 - E_G} \tag{22}$$

where w_{SG} and w_{SL} are the superficial gas and liquid velocities in the column.

The effective velocity of the liquid phase w_L is given by Eq. (23)

$$w_L = \frac{w_{SL}}{1 - E_G} \tag{23}$$

and controlled by the distribution of residence times. For the estimation of the coefficient of longitudinal dispersion and/or back mixing two different methods were used: a nonstationary and/or a stationary method. The nonstationary method was carried out by a pulse tracer technique using a 20% NaCl solution. The tracer was injected into the column through a small stainless steel tube. The opening time of the magnetic valve was varied between 0.3 and 1.0 sec. The concentration of tracer in the column was measured at two locations by means of conductivity cells installed at a distance of $L = 100$ cm from each other in the center of the column to avoid inlet and outlet effects. This tracer technique has the advantage that a mathematically perfect pulse is unnecessary. The tracer concentration was measured by conductivity detectors in the column and recorded by data logger on paper tape as a discrete function of time (for more details see [36]).

The stationary method involved the use of dye-solution and/or sodium chloride solution as tracer approximating a plane source. The test section of the column contained 19 radial openings for sampling, each at an axial distance of 5 cm from the other. Stainless steel tubes of 5 mm outer diameter were used for sampling, each capable of being moved radially from the column wall to its center. The concentrations of the dye and/or of the sodium chloride in the liquid phase were measured by photometer and/or a conductivity transmitter (Philips type PR 9507) connected to a millivolt meter and an electrode (Philips type PW 9513). The longitudinal concentration profile of the tracer was determined upstream of the source.

The bubble size distribution was measured photographically and/or by the electro-optical method of Todtenhaupt [58] and Pilhofer [59]. To aid the photographic measurements plane parallel windows were attached to the column to avoid optical distortion. Two different exposure conditions were used: for large bubbles aperture 22, exposure time 1/1000 sec and for small bubbles aperture 8, exposure time 1/5000–1/4000 sec (for further details see [60]).

4. Applied Mathematical Models

To estimate the model parameter, *Pe*, of the axial dispersion model [Eq. (21)] the variances of the experimentally determined residence time distributions were evaluated at two axial positions. The difference of these variances was used as a first approximation of the parameters as a basis for the further calculation. The system transfer function for the axial dispersion model [37] was expressed as:

$$F(s^*, Pe) = \exp\left[\frac{Pe}{2}\left(1 - \sqrt{1 + \frac{4\,s^*}{Pe}}\right)\right]. \tag{24}$$

The experimental transfer functions were evaluated from the experimental residence times distribution by numerical integration of the transient response of an imperfect pulse:

$$F_M(s^*) = C_2^*(s^*)/C_1^*(s^*). \tag{25}$$

In Eqs. (24) and (25) F and F_M are the calculated and measured transfer functions, s^* the dimensionless Laplace variable, $\overline{C_1^*}(s^*)$ and/or $\overline{C_2^*}(s^*)$ the Laplace transform of $C_1(t)$ and/or $C_2(t)$.

The model parameters were estimated by the least-squares error analysis of the transformed response data, i.e.

$$\phi(P_j, s_i^*) = \sum_{i=1}^{n} [F_M(s_i^*) - F(s_i^*, P_j)]^2 \tag{26}$$

The analysis requires that the function $\phi(P_j, s_i^*)$ is minimized with respect to the model parameters P_j ($j = 1, 2, ..., m$) for selected values of the dimensionless Laplace variable $s^* = s \cdot \bar{t}$, where t = the mean residence time. For the determination of the least squares estimates the methods of Gauss-Seidel [61] and Marquardt [62] were applied. The standarf error SE of the estimate is calculated as:

$$SE = \sqrt{\frac{\phi(P_j, s_i^*)}{n - m}}$$

where n is the number of data points used for the fit and m the number of parameters. The model and experimental transfer functions are in good agreement. The standard deviation of the fitted Peclet number is $SE = 2.5 \cdot 10^{-3}$ (for details see [36]).

The longitudinal dispersion coefficient was estimated from the data obtained using the stationary method by means of the balance equation:

$$w_L \frac{dC}{dz} = D_L \frac{d^2C}{dz^2}, \quad (27a)$$

where C the concentration of the tracer

z the *upstream* longitudinal distance from the source.

With the boundary condition

$$C = C_A \text{ for } z = 0$$

one obtains the solution for $z < 0$

$$\frac{C}{C_A} = \exp\left(-\frac{w_L \cdot z}{D_L}\right). \quad (27b)$$

By fitting the calculated longitudinal profiles to the measured ones D_L was estimated. Since D_L was evaluated from the longitudinal concentration profile of the tracer upstream from the source it is called the coefficient of backmixing D_{LB} in contrast to $D_{L\,eff}$. Since the latter was estimated by the nonstationary method from the distribution of residence times, DRT, it is a dispersion coefficient which, according to G. I. Taylor [101], does not necessarily have a feed-back effect, while D_{LB} represents true backmixing with feed-back. The relation between these mixing coefficients is discussed in [102].

A two-phase longitudinal dispersion model, which considers the linear change of the gas pressure along the column [63], was used for the estimation of the volumetric mass transfer coefficient ($k_L a$). The liquid phase mass balance is given by Eq. (28):

$$\frac{d^2c_L}{dx^2} - \frac{dc_L}{dx} \cdot Bo + St \cdot Bo \cdot \left[\frac{y_{Go}}{He}(g \cdot \rho_L \cdot H_B(1 - E_G)(1 - x) + P_0) - c_L\right] = 0. \quad (28)$$

The boundary conditions are at the *downstream* side of the gas entrance (gas distributor):

at $X = 0^{(+)}$ $$C^* - \frac{1}{Bo}\left(\frac{dC^*}{dx}\right) = 0 \quad (28a)$$

and at the *upstream* side of the exit of both of the phases:

$$\text{at} \qquad X = 1^{(-)} \qquad \frac{dC^*}{dx} = 0. \tag{28b}$$

Here $X = \frac{x}{L}$ the dimensionless longitudinal coordinate and

$C^* = \frac{C_L}{C_{LG}} = \frac{C_L}{C_L^*}$ the dimensionless concentration with regard to the saturation concentration C_{LG}.

It is necessary to apply C^* in the boundary conditions to eliminate the influence of the longitudinal pressure variation which causes a finite gradient dc_L/dx at the exit (see Figs. 3–5). At the top of the bubbling layer $X = 1^{(+)}$, atmospheric pressure prevails which does not depend on X for $X \geqslant 1$. Therefore at the *downstream* side of the exit:

$$X = 1^{(+)} \qquad \frac{dC^*}{dx} = 0 \quad \text{and} \quad \frac{dC_L}{dx} = 0.$$

By means of these so called "Danckwert's boundary conditions" (28) the solution is given by [64]:

$$c_L = B_4 \cdot e^{\frac{Bo}{2}(1+q)^x} + \overline{B}_5 \cdot e^{\frac{Bo}{2}(1-q)^x} + \overline{B}_2 \cdot x + \overline{B}_1 \tag{29}$$

with
$$Bo = \frac{w_L \cdot H}{D_L},$$
$$St = k_L \cdot a\,\bar{t}$$
$$q = \sqrt{1 + 4\,St/Bo}$$
$$\overline{B}_1 = \alpha - \beta/St$$
$$\overline{B}_2 = \beta$$
$$\overline{B}_4 = D_1/D_0$$
$$B_5 = D_2/D_0$$
$$\alpha = \frac{y_{GO}}{He}\left(1 + \frac{g \cdot \rho_L \cdot E_L \cdot H_B}{P_0}\right)$$
$$\beta = \frac{y_{GO}}{He} \cdot \frac{g\,\rho_L}{P_0} \cdot E_L \cdot H_B$$
$$D_0 = Bo(1-q)^2 \cdot e^{\frac{Bo}{2}(1-q)} - Bo\,(1+q)^2 \cdot e^{\frac{Bo}{2}(1+q)}$$
$$D_1 = Bo\,(1-q) \cdot 2 \cdot \left(\frac{\overline{B}_2}{Bo} - \overline{B}_1\right) e^{\frac{Bo}{2}(1-q)} + 2\,(1+q) \cdot \overline{B}_2$$

and
$$D_2 = -2\,(1-q) \cdot \overline{B}_2 - 2\,Bo\,(1+q)\left(\frac{\overline{B}_2}{Bo} - \overline{B}_1\right) e^{\frac{Bo}{2}(1+q)}.$$

The longitudinal concentration profile of oxygen in the liquid phase can be described by Eq. (29) if one assumes a linear pressure drop gradient along the column.
In Figs. 2 and 3 calculated longitudinal concentration profiles of oxygen in the liquid phase are plotted as function of *Bo* and *St* numbers.

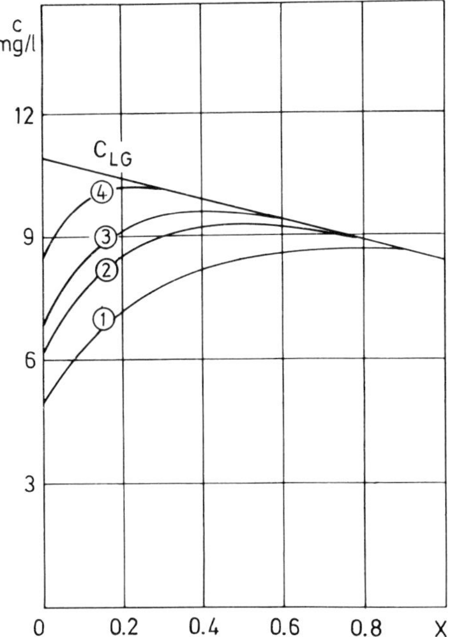

Fig. 2. Calculated longitudinal concentration profiles of oxygen for Bo = 3 and different St-numbers. E_G = 0.2

C_{LG} concentration of dissolved oxygen in equilibrium with the gas phase (saturation concentration)

$X = \frac{x}{L}$ dimensionless longitudinal coordinate. X = 0 (gas entrance) X = 1 (exit)

(1) $C_L(X)$ for St = 5
(2) $C_L(X)$ for St = 10
(3) $C_L(X)$ for St = 15
(4) $C_L(X)$ for St = 50

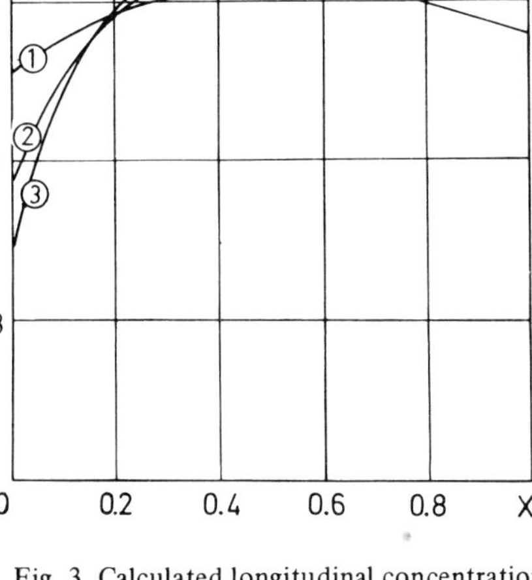

Fig. 3. Calculated longitudinal concentration profiles of oxygen for St = 10 and different Bo-numbers. E_G = 0.1
(1) $C_L(X)$ for Bo = 1
(2) $C_L(X)$ for Bo = 5
(3) $C_L(X)$ for Bo = 10
(for symbols see Fig. 2)

On these figures C_{LG} = the saturation concentration, C_L = the concentration of the dissolved gas (O_2). X is the dimensionless longitudinal position in the column. At $X = 0$ the gas enters into the system and at $X = 1$ both of the phases leave the column. Because of the linear longitudinal pressure drop gradient the saturation concentration, C_{LG}, diminishes linearly along the column. The smaller the Stanton number St the later the saturation is achieved. With $St = 5$ the liquid is sautrated at $X = 0.9$. With increasing St this saturation position is shifted to lower X's. With $St = 50$ the liquid is saturated at $X = 0.3$ already. The St number has significant influence on $C_L(X)$ (Fig. 2). At low Bodenstein number, Bo, the concentration profiles are flat because of the high intensity of longitudinal dispersion. With increasing Bo the profiles become steeper. However, the Bo-number has a relatively slight influence on these profiles (Fig. 3). The fitting of the calculated concentration profiles to the measured ones was carried out by non-linear regression using the optimisation method of Gauss-Seidel [61]. The normalized concentrations (with regard to their equilibrium value) are used for this fitting. (For details see [57].) The standard errors *SE*

$$SE = \sqrt{\frac{\phi(Bo, St)}{K - Pi}} = \sqrt{\frac{\sum_1^K [C^*_{GMN} - C_{LN}(Bo, St)]}{K - Pi}} \tag{30}$$

where K = the number of points,
Pi = the number of parameters,

are always lower than 0.015. Typical calculated and measured longitudinal profiles are shown in Figs. 4 and 5. All of the profiles are calculated on the assumption that the partial pressure of O_2 in the gas phase along the column is constant. Because of the slight change of this partial pressure (< 0.02 atm) this assumption is a realistic one.

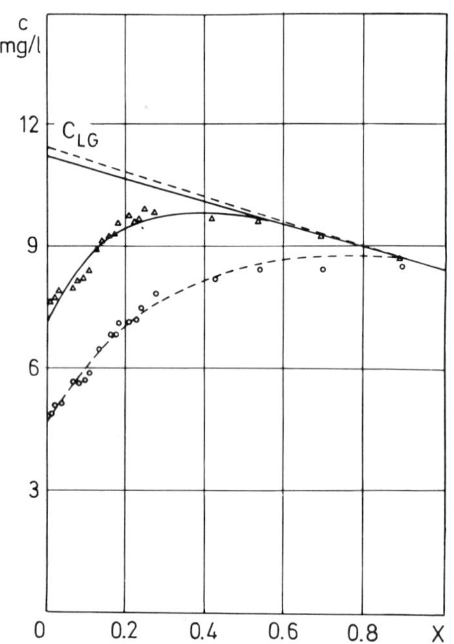

Fig. 4. Typical calculated and measured longitudinal concentration profiles. Perforated plate, H_2O, w_{SL} = 2.2 cm/s

□	w_{SG} = 1.61 cm/s	E_G = 0.06
---	Bo = 3.39	St = 4.35
△	w_{SG} = 3.22 cm/s	E_G = 0.118
——	Bo = 3.02	St = 15.68

(for other symbols see Fig. 2)

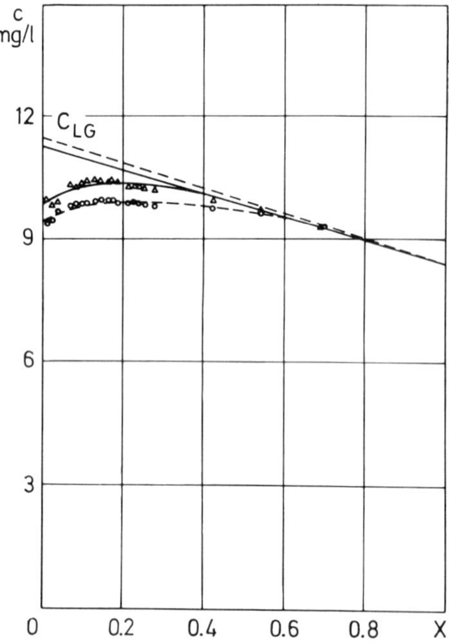

Fig. 5. Typical calculated and measured longitudinal concentration profiles. Porous plate, 1% CH_3OH/NMP, w_{SL} = 2.2 cm/s

□	w_{SG} = 0.81 cm/s	E_G = 0.038
---	Bo = 0.40	St = 28.7
△	w_{SG} = 1.88 cm/s	E_G = 0.10
——	Bo = 0.72	St = 69.9

(for other symbols see Fig. 2)

5. Experimental Comparison of Single-Stage Bubble Columns with Different Aerator Types and Fermentation Media

a) Mean Relative Gas Hold-Up E_G

It is well known that the most important difference between pure liquid and (antifoam free) solutions of salts and C_1–C_4 alcohols and acids is that in the former the bubble coalescence rate is high and in the latter the coalescence rate is low [31, 32, 50, 57, 63, 65–68].

Demineralized water was used as a reference system to compare fermentation media with systems of high coalescence rate. On the other hand, in high concentration salt solutions the coalescence rate is strongly diminished. Therefore 10% Na_2SO_4 solution was used as second reference solution. This solution corresponds to the sodium sulphite solution which is often used to estimate gas/liquid interfacial areas by means of mass transfer limited chemical reaction. The application of this solution makes it possible to compare the systems investigated with systems studies by other groups.

The aqueous solutions used contained methanol, ethanol, n-propanol, n-butanol and glucose in concentration ranges 0.5–20% without and with salt (ca. 1%), while perforated plate, porous plate, injector nozzle and ejector nozzle aerators all received attention.

In bubble columns with pure water no differences can be found between the gas hold-ups with porous and perforated plates. E_G increases with the gas flow rate w_{SG} but it is independent of the liquid flow rate w_{SL} (Fig. 6). With the injector nozzle and for

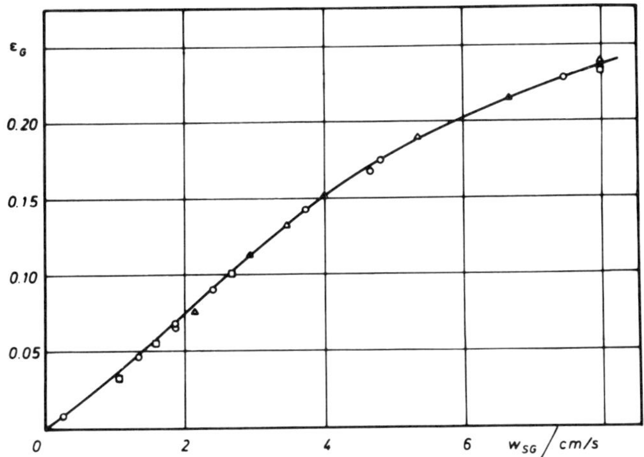

Fig. 6. Influence of the gas distributor type on E_G in demineralized water.

○ w_{SL} = 1.21 cm/s, □ w_{SL} = 2.20 cm/s } porous plate

△ w_{SL} = 1.69 cm/s, ▲ w_{SL} = 2.20 cm/s } perforated plate

w_{SG} superficial gas velocity [cm/s]

w_{SL} superficial liquid velocity [cm/s]

The units of the coordinates are given according to the German standard, i.e. w_{SG}/cm/s means w_{SG} [cm/s].

$w_{SG} < 3$ cm/s the gas hold-up is independent of w_{SL} (Fig. 7). At higher gas flow rates E_G depends on w_{SL} and on the quality of the water. With increasing w_{SG} the gas hold-up passes a maximum. The higher w_{SL} and the lower the salt concentration the lower is the maximum, since with increasing w_{SL} and decreasing salt concentration the coalescence rate increases. The behaviour of the ejector nozzle is similar to that of the injector nozzle (Fig. 8). A comparison of the gas hold-ups with the four aerator types shows that at low w_{SG} the aerator exerts only weak influence on E_G. At higher gas

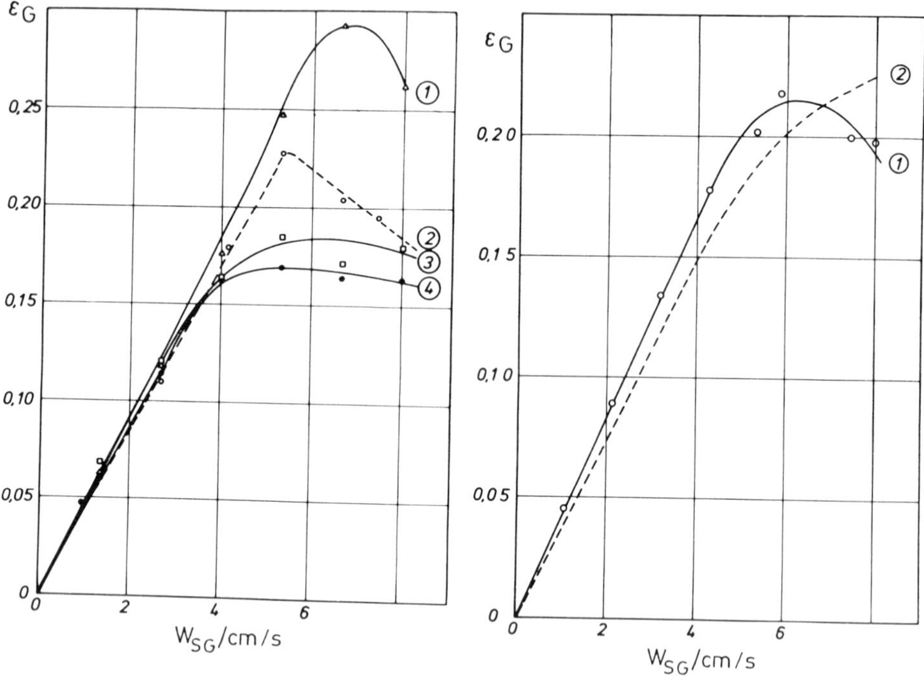

Fig. 7. Gas hold up as function of the superficial gas and liquid velocity, injector nozzle – with water.
(1) tap water w_{SL} = 1.2 cm/s
(2) demineralized water w_{SL} = 1.2 cm/s
(3) demineralized water w_{SL} = 1.8 cm/s
(4) demineralized water w_{SL} = 2.2 cm/s

Fig. 8. Gas hold up as function of the superficial gas velocity. Demineralized water.
(1) ejector nozzle
w_{SL} = 1.46 cm/s
(2) porous- and perforated plate
w_{SL} = 1.21–2.2 cm/s

flow rates the E_G (w_{SG}) curve passes through a maximum only with nozzle aerators in the range of w_{SG} investigated. The size of the bubbles formed near to a nozzle, is controlled by the dynamic equilibrium diameter $d_{B\max}$ according to Eq. (18), consequently the size increases with growing distance from the nozzle due to decreasing energy dissipation density. This dynamic equilibrium diameter is attained in systems with a high coalescence rate and with a nozzle aerator earlier than with a porous or perforated plate, since in systems with a nozzle the coalescence is accelerated by the strongly non-uniform velocity distribution of the liquid. It is well known [69] that in the presence of a non-uniform velocity profile with a velocity maximum in the center the large bubbles are driven into the center of the column due to the higher flow resistance acting on the bubble surface area in greater distance from the center. The enrichment of large bubbles in the column center increases the non-uniformity of the transverse liquid flow velocity. Large bubbles are very effective in collecting the small bubbles in their wake, since the small bubbles are accelerated in the direction of the large bubbles due to the diminished flow resistance in the wake and are sucked into the large bubbles. Increased number density of bubbles in the column center increases the coalescence rate because more bubbles come into a position favourable for coalescence. This increases the coalescence rate systems with ejector and injector nozzles and causes the earlier maximum of the $E_G(w_{SG})$ curve and the lower E_G values at high gas flow rates in comparison with the flat gas distributors.

Since the liquid flow rate has only a slight effect on E_G in the following the liquid flow rate is not explicitly considered. Depending on the type of the gas distributor the salt and substrate effects can be very different (Fig. 9). The addition of salt and substrates

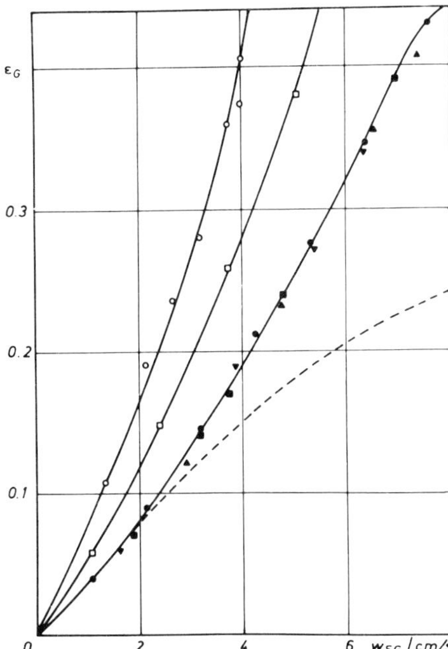

Fig. 9. Typical influence of substrate and salt additives on E_G. w_{SL} = 1.21 cm/s

Porous plate	perforated plate
○ 1% C_2H_5OH + NMP	● 1% C_2H_5OH + NMP
□ 1% CH_3OH + NMP	■ 1% CH_3OH + NMP
-- H_2O	▲ 1% C_2H_5OH
	▼ 1% CH_3OH
	-- H_2O

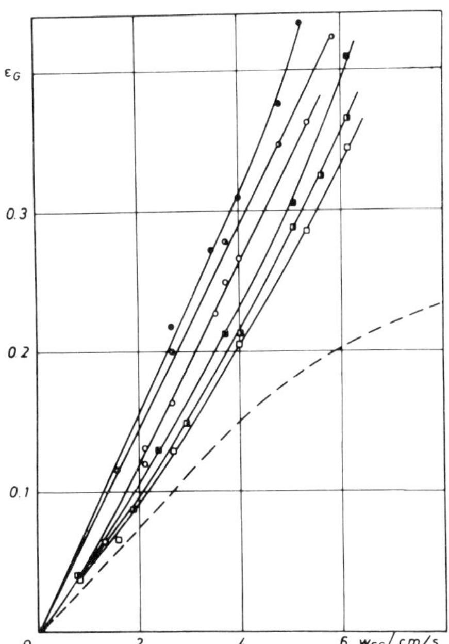

Fig. 10. Typical influence of the concentration of methanol and ethanol on E_G.
Porous plate. w_{SL} = 1.21 cm/s

○ 0.5% C_2H_5OH	□ 0.5% CH_3OH
◑ 1% C_2H_5OH	◨ 1% CH_3OH
● 2% C_2H_5OH	■ 2% CH_3OH
-- H_2O	

to the water phase causes a non-specific increase of E_G for the perforated plate and a specific increase for the porous plate. For the latter the variation of E_G depends on the composition of salt additives and on the type of solute and its concentration (Figs. 9, 10, and 11). At the same gas flow rate the addition of C_2H_5OH produces higher E_G values than CH_3OH. With increasing concentration of alcohols, E_G also increases without salts (Fig. 10) as well as with salts (Fig. 11), as long as the alcohol concentration is low enough. However, with further increase in the concentration of alcohols E_G passes a maximum. The position of this maximum shifts with increasing chain length of the alcohol solute to lower concentrations (Fig. 12). E_G increases with gas flow rate w_{SG} for all of the systems investigated; the sensitivity of E_G to a change in w_{SG} depends on the additives (Figs. 9–11).

Since alcohol concentrations higher than 1% are not interesting for fermentation, the $E_G(w_{SG})$ curves for the concentration range between 1 and 20% are not shown.

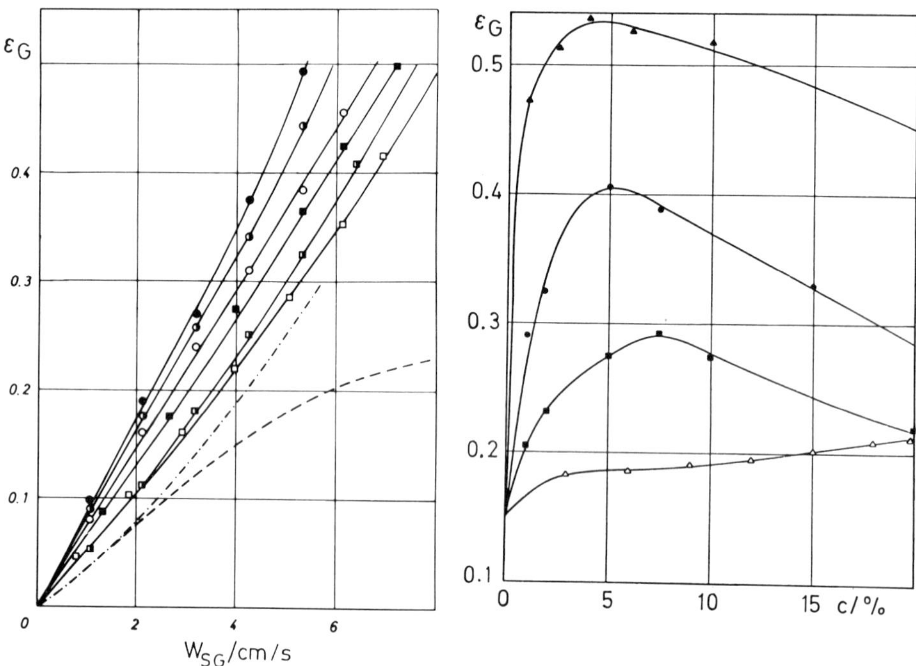

Fig. 11. Typical influence of the concentration of methanol and ethanol with salt additives on E_G. Porous plate. w_{SL} = 2.2 cm/s

□ 0.5% CH_3OH/NOP ○ 0.5% C_2H_5OH/NOP
◨ 1% CH_3OH/NOP ◑ 1% C_2H_5OH/NOP
■ 2% CH_3OH/NOP ● 2% C_2H_5OH/NOP
–·–·– H_2O/NOP – – – H_2O

Fig. 12. Typical influence of the concentration of solute on E_G. Porous plate. w_{SL} = 2.2 cm/s, w_{SG} = 4.0 cm/s

■ CH_3OH ● C_2H_5OH
▲ n-C_3H_7OH △ glucose

However, it is worth mentioning that, for example, when w_{SG} = 6.5 cm/s and w_{SL} = 2.2 cm/s very high gas hold-ups can be achieved: i.e.

E_G = 0.58 with 7.5% CH_3OH
E_G = 0.62 with 5.0% C_2H_5OH
E_G = 0.66 with 1.0% n-C_3H_7OH and
E_G = 0.64 with 1.0% n-C_4H_9OH.

The presence of yeast cells (*Candida boidinii*) does not have a great influence on E_G, as long as their concentration is low: with the perforated plate no difference was found between the systems without and with cells at all (Fig. 13). Also with the porous plate this influence is slight (Figs. 14 and 15). These results were confirmed by measurements carried out in bubble column fermentors during cell growth [70].

The type of solute and its concentration influences the hold-up systems with the injector nozzle in a manner similar to that obtained with the porous plate. Again methanol produced the smallest increase of E_G (Fig. 16), because its hinderance on the coalescence

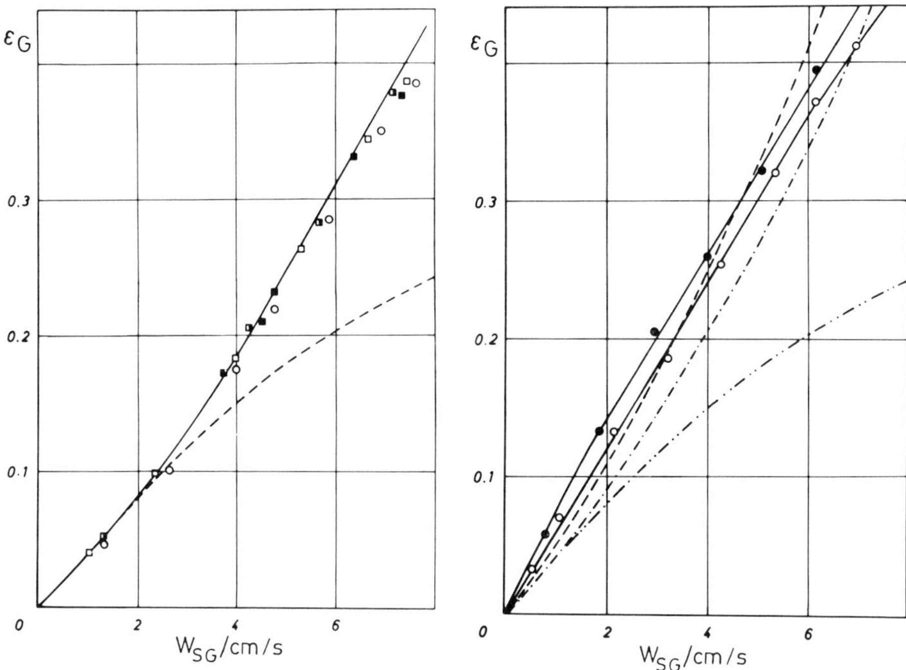

Fig. 13. Typical influence of cell concentration on E_G. Perforated plate. w_{SL} = 1.69 cm/s
□ 0.3% yeast
◧ 1% CH_3OH/NOP, 1.0% yeast
■ 1.5% yeast
○ H_2O/NOP 1.5% yeast
—— 1% CH_3OH/NOP
– – – H_2O

Fig. 14. Typical influence of cell concentration on E_G. Porous plate/methanol system. w_{SL} = 2.2 cm/s
● 1% CH_3OH/NMP/0.3% yeast
– – – 1% CH_3OH/NMP
○ 0.5% CH_3OH/0.6% yeast
–.–.– 0.5% CH_3OH
–..–. H_2O

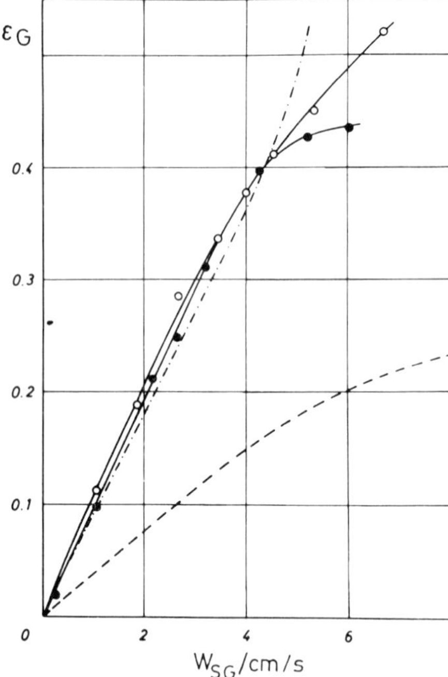

Fig. 15. Typical influence of cell concentration E_G. Porous plate/ethanol system.
w_{SL} = 1.21 cm/s
○ 1% CHOH/NOP/0.3% yeast
● 1% C_2H_5OH/NOP/0.6% yeast
–·–·– 1% C_2H_5OH/NOP
––– H_2O

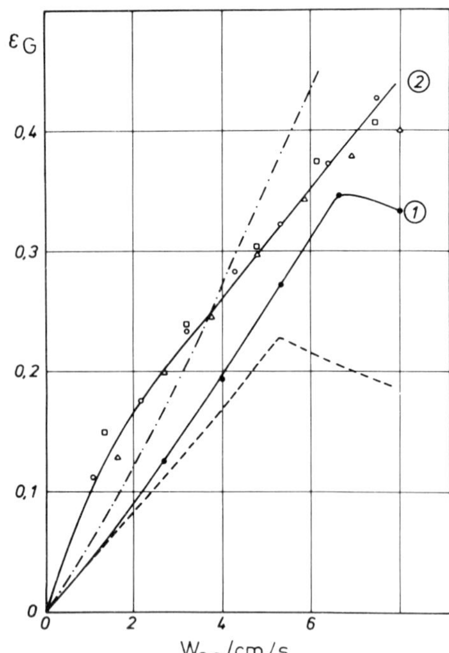

Fig. 16. Gas hold up as function of the superficial gas velocity. Injector nozzle–methanol system.

(1) 1% CH_3OH w_{SL} = 1.2 cm/s
(2) ○ 1% CH_3OH w_{SL} = 2.2 cm/s
□ 1% CH_3OH/NMP w_{SL} = 1.2 cm/s
△ 1% CH_3OH/NMP w_{SL} = 2.2 cm/s
––– H_2O, w_{SL} = 1.2 cm/s
–·–·– porous plate, 1% CH_3OH/NMP, w_{SL} = 1.6 cm/s

is the smallest, n-propanol (Fig. 18) and n-butanol (Fig. 19) the highest, and ethanol (Fig. 17) produced an intermediate increase in E_G. For ethanol the injector nozzle is better than porous plate, for n-propanol and n-butanol they are comparable. It seems that the injector nozzle was not fully optimized with regard to the liquid flow rate, since it would be expected to give a higher gas hold-up for propanol and butanol than the porous plate. A comparison of the alcohol- and sulphate-systems indicates that the

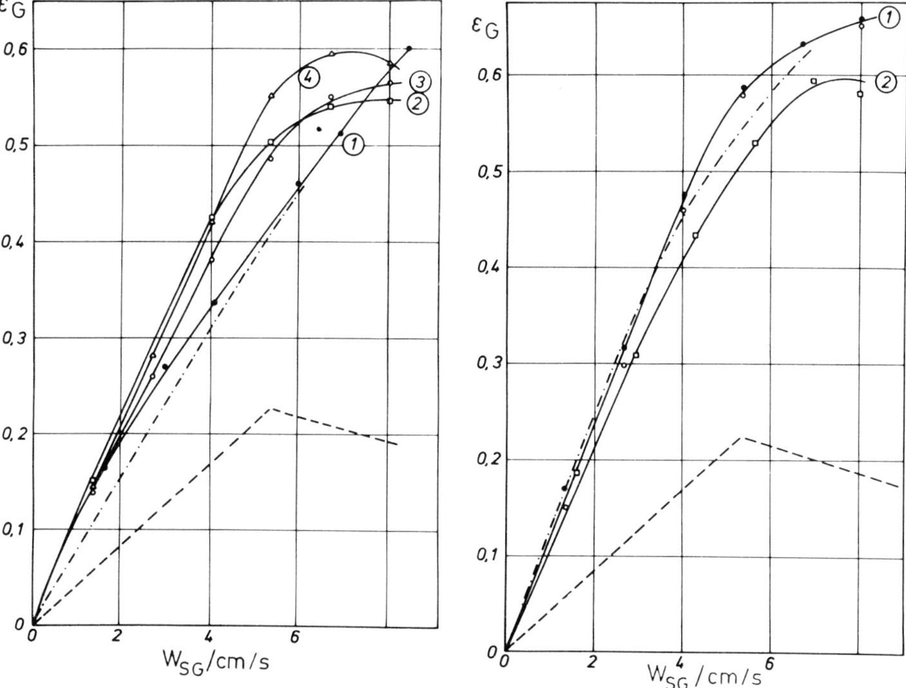

Fig. 17. Gas hold up as function of the superficial gas velocity. Injector nozzle–ethanol system.
(1) 1% C_2H_5OH w_{SL} = 1.8 cm/s
(2) 1% C_2H_5OH w_{SL} = 2.2 cm/s
(3) 1% C_2H_5OH w_{SL} = 2.56 cm/s
(4) 1% C_2H_5OH + NMP, w_{SL} = 2.2 cm/s
––– H_2O w_{SL} = 1.2 cm/s
–·–·– porous plate, 1% C_2H_5OH/NMP, w_{SL} = 2.2 cm/s

Fig. 18. Gas hold up as function of the superficial gas velocity. Injector nozzle–1% n-propanol system.
(1) { • w_{SL} = 1.8 cm/s; ○ w_{SL} = 2.2 cm/s }
(2) □ w_{SL} = 2.56 cm/s
––– H_2O w_{SL} = 1.2 cm/s
–·–·– porous plate w_{SL} = 2.2 cm/s

behaviour of bubble columns with n-propanol, n-butanol and/or Na_2SO_4 is comparable (Fig. 20). At high gas velocities the $E_G(w_{SG})$ curve for sodium sulphate approaches a constant value earlier than propanol or butanol. This is either due to a stronger anti-coalescence effect by these alcohols compared with sodium sulphate, or due to the high sensitivity of the injector nozzle to the liquid flow rate. A significant maximum of E_G as function of w_{SL} prevails for the injector nozzle. The position of this maximum

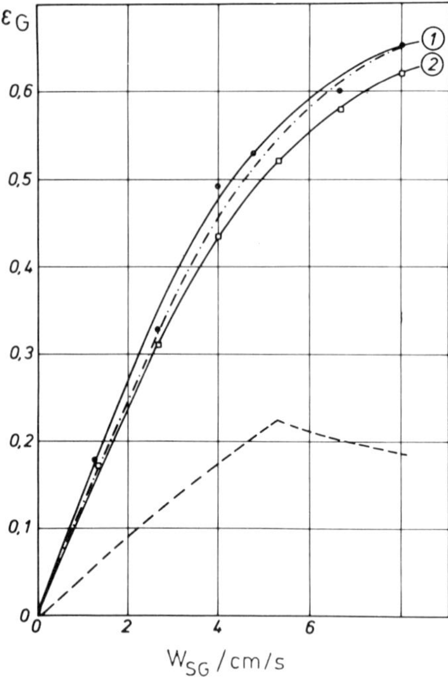

Fig. 19. Gas hold up as function of the superficial gas velocity. Injector nozzle – 1% n-butanol system.
(1) w_{SL} = 1.8 cm/s
(2) w_{SL} = 2.2 cm/s
– – – H_2O, w_{SL} = 1.2 cm/s
–·–·– porous plate, w_{SL} = 2.2 cm/s

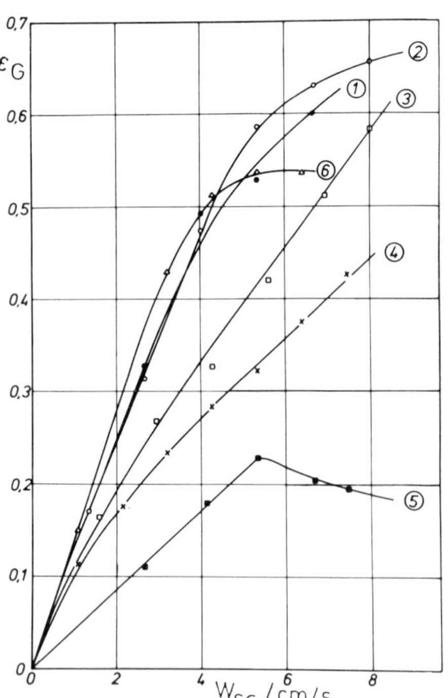

Fig. 20. Gas hold up as function of the superficial gas velocity. Injector nozzle. Comparison of the alcohol and sulphate systems.
(1) 1% n-butanol w_{SL} = 1.8 cm/s
(2) 1% n-propanol w_{SL} = 1.8 cm/s
(3) 1% ethanol w_{SL} = 1.8 cm/s
(4) 1% methanol w_{SL} = 1.8 cm/s
(5) H_2O w_{SL} = 1.2 cm/s
(6) 10% Na_2SO_4 w_{SL} = 1.8 cm/s

depends on the composition of the liquid. It is possible that, because of the stepwise change of w_{SL} during the experimentation, the exact optimum of w_{SL} was not found for the 10% Na_2SO_4 solution. The ejector nozzle is not as sensitive to the liquid flow rate as the injector nozzle. This may be the explanation for the fact that the ejector nozzle produces high gas hold-ups for high flow rates of sodium sulphate solution (Fig. 22) in contrast to the injector nozzle. Disregarding this difference the behaviour

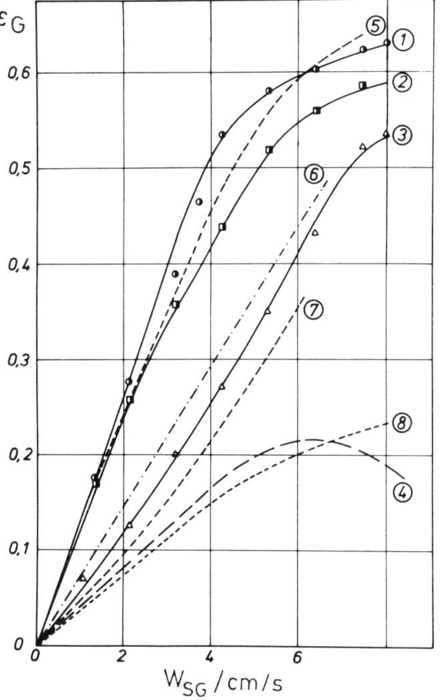

Fig. 21. Gas hold up as function of the superficial gas velocity. Comparison of ejector nozzle with porous plate for 1% alcohol systems.

(1) propanol, (2) ethanol, (3) methanol, (4) water — ejector nozzle, w_{SL} = 1.5 cm/s

(5) propanol, (6) ethanol, (7) methanol, (8) water — porous plate, w_{SL} = 1.2 cm/s

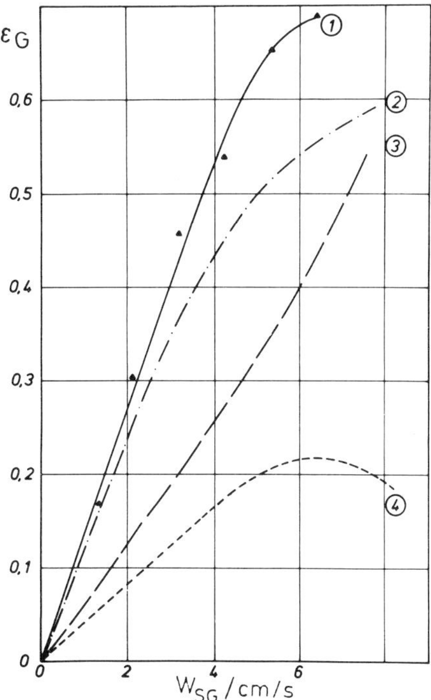

Fig. 22. Gas hold up as function of the superficial gas velocity. Ejector-nozzle. Comparison of alcohol and sulphate systems.

(1) 10% Na_2SO_4 w_{SL} = 1.66 cm/s
(2) 1% C_2H_5OH w_{SL} = 1.58 cm/s
(3) 1% CH_3OH w_{SL} = 1.50 cm/s
(4) H_2O w_{SL} = 1.2–2.2 cm/s

of these two nozzles is very similar. With the ejector nozzle E_G increases as follows: water < methanol < ethanol < propanol (Fig. 21). The differences between the ejector nozzle and the porous plate are relatively slight when using methanol and propanol, but (similar to the injector nozzle) fairly large for ethanol. This behavior is not yet understood.

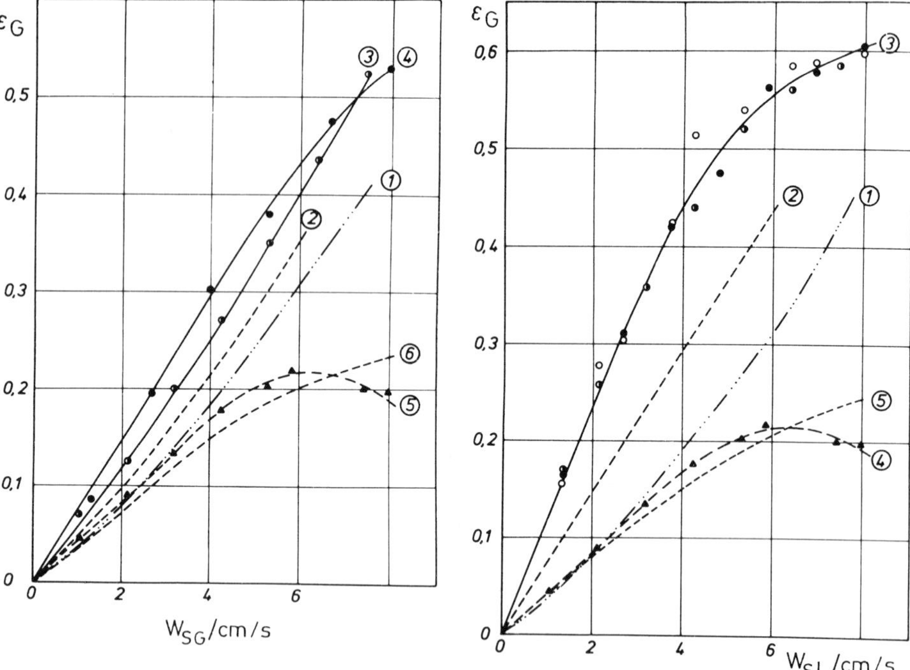

Fig. 23. Gas hold up as function of the superficial gas velocity. Methanol system. Comparison of different aerator types.
(1) perforated plate: 1% CH_3OH, w_{SL} = 2.2 cm/s
(2) porous plate: 1% CH_3OH, w_{SL} = 1.21 cm/s
(3) ejector nozzles: 1% CH_3OH, w_{SL} = 1.5 cm/s
(4) ejector nozzle: 2% CH_3OH, w_{SL} = 1.5 cm/s
(5) ejector nozzle: water, w_{SL} = 1.5 cm/s
(6) perforated plate and porous plate: water, w_{SL} = 2.2 cm/s

Fig. 24. Gas hold up as function of the superficial gas velocity. Ethanol system. Comparison of different aerator types.
(1) perforated plate: 1% C_2H_5OH w_{SL} = 2.2 cm/s
(2) porous plate: 2% C_2H_5OH w_{SL} = 1.2 cm/s
(3) ejector nozzle: w_{SL} = 1.52 cm/s
○ 0.5% C_2H_5OH, ○ 1% C_2H_5OH
● 2.0% C_2H_5OH
(4) ejector nozzle: H_2O
(5) perforated plate and porous plate: H_2O

The ejector nozzle gives, in general, a slightly higher gas hold-up than the injector nozzle. A comparison of gas hold-up values for the different aerator types in methanol (Fig. 23), ethanol (Fig. 24), ethanol + salt (Fig. 25), alcohol + salt (Fig. 26) as well as sodium sulphate solutions (Fig. 28) indicates that the ejector nozzle is more effective than the porous plate while the perforated plate always produces the smallest gas hold-up.
The glucose system is an exception (Fig. 27). For $w_{SG} < 5$ cm/s no significant difference in gas hold-up can be recognized. Above this gas flow rate the porous plate

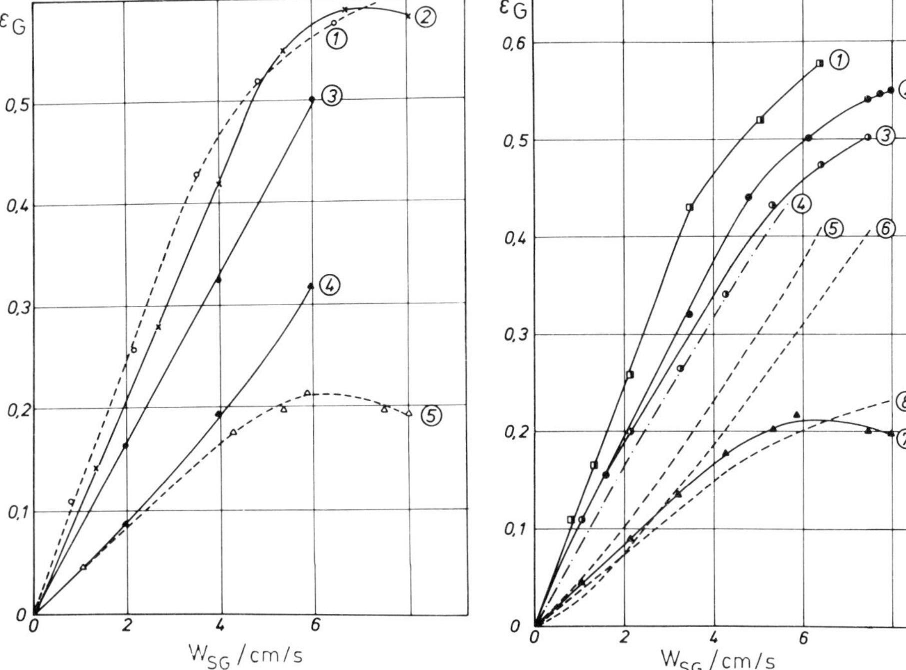

Fig. 25. Gas hold up as function of the superficial gas velocity. 1% ethanol + 1% salt system. w_{SL} = 1.5–2.2 cm/s. Comparison of different aerator types.

(1) ejector nozzle
(2) injector nozzle
(3) porous plate
(4) perforated plate
(5) ejector nozzle: H_2O

Fig. 26. Gas hold up as function of the superficial gas velocity. Alcohol-salt-systems. Comparison of different aerator types.

(1) ejector nozzle: 1% C_2H_5OH
(2) ejector nozzle: 2% C_2H_5OH } w_{SL} = 1.5 cm/s
(3) ejector nozzle: 1% CH_3OH
(4) porous plate: 1% C_2H_5OH
(5) porous plate: 1% CH_3OH } w_{SL} = 2.2 cm/s
(6) perforated plate: 1% C_2H_5OH
(7) ejector nozzle: H_2O
(8) porous plate: H_2O

produces a higher E_G than the ejector nozzle. It seems that the coalescence in glucose solutions is influenced only slightly. Therefore the difference in E_G for pure water and glucose solution is small.

According to these results the gas hold-up increases with different aerator types in following sequence: perforated plate < porous plate, injector nozzle < ejector nozzle, with the exception of solutions with a high and/or medium coalescence rate (water and/or glucose solutions). E_G also increases with different substrates and/or salt solutions in the sequence: demineralized water < 1% salt solution < 1% glucose solution < 0.5–2% methanol solution < 0.5–2% methanol + 1% salt solution < 0.5–2% ethanol solution, 0.5–2% ethanol + 1% salt solution < 1% n-propanol-solution, 1% n-butanol-solution, 10% Na_2SO_4 solution with exception of the perforated plate, for which the composition of the liquid has no significant effect on E_G.

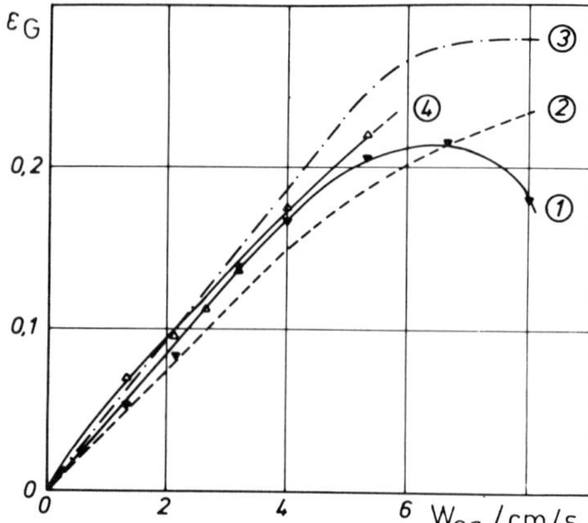

Fig. 27. Gas hold up as function of the superficial gas velocity. Glucose-system. Comparison of different aerator types.
(1) ejector nozzle: 1% glucose
(2) porous plate: H_2O
(3) porous plate: 1% glucose
(4) injector nozzle: 1% glucose

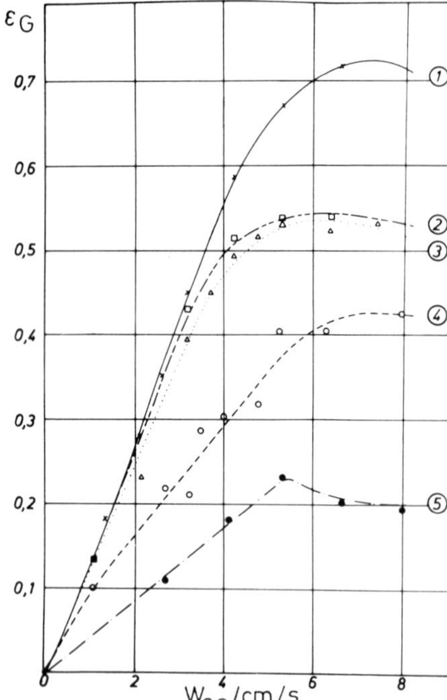

Fig. 28. Gas hold up as function of the superficial gas velocity. 10% Na_2SO_4 system. Comparison of different aerator types.
(1) ejector nozzle: w_{SL} = 1.5 cm/s
(2) injector nozzle: w_{SL} = 1.8 cm/s
(3) injector nozzle: w_{SL} = 2.2 cm/s
(4) porous plate: w_{SL} = 2 2 cm/s
(5) injector nozzle: w_{SL} = 1.2 cm/s
(in water)

b) Relation for the Mean Relative Gas Hold-Up E_G

A large number of measurements of E_G, carried out in bubble columns, are reported in the literature. While most of these investigations only involved a few aspects, some of them resulted in general correlations between E_G and the most important parameters for pure liquids. The experimental results of the present study for water were therefore compared with data calculated by applying the general equations of Hughmark [71], Akita [66], Burkel [72] and Gestrich [51]. In these equations only the direct proportionality of E_G to w_{SG} is common. The hole diameter of the gas distributor and the properties of the liquid phase are considered in various ways.

Whereas the experimental results for water with the perforated plate gas distributor can be well described by the equations of Akita [66] and Gestrich [51], none of these equations is able to yield relations in agreement with the experimental results obtained when using either the porous plate and/or the ejector and/or injector nozzle gas distributors. An attempt to change the above equations by varying the influence of the pore and/or hole diameter of the gas distributor and the pressure drop across it to take care of the relative gas hold-up of these systems did not succeed, i.e. the equations were unable to describe the behaviour of bubble columns with the porous plate and the nozzle gas distributors in the investigated range of gas and liquid flow rates for pure water. Because of the eminent role of the bubble size which can be recognized from the results of the present paper, the Sauter mean bubble diameter was used to describe the mean relative gas hold-up. By using the superficial gas velocity w_{SG} and the Sauter mean diameter d_s the following relation was obtained:

$$\mathrm{E}_G = 0.91\, Fr^{1.19} \tag{31}$$

where the Froude number, Fr, is defined as

$$Fr = \frac{w_{SG}}{\sqrt{g d_s}} \, .$$

The measured mean relative gas hold-ups are plotted in Fig. 29 for the various systems investigated, superimposed on this figure are data calculated according to Eq. (31). The agreement is satisfactory with a mean relative error of 14.5%.

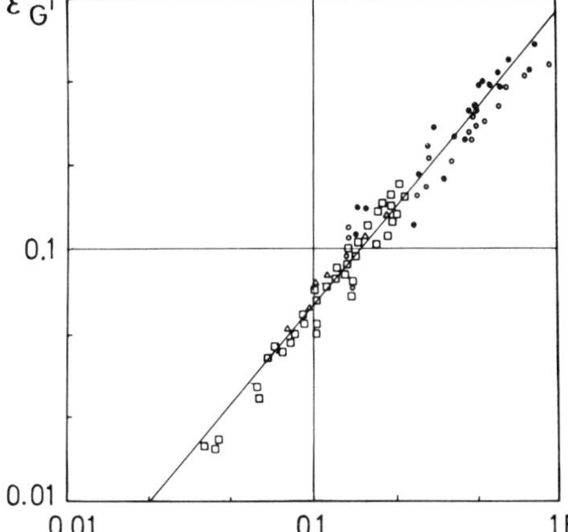

Fig. 29. Comparison of the measured gas hold up as function of the Froude number with the data calculated using Eq. (31).
- ● ejector nozzle
- ○ injector nozzle
- □ porous plate
- △ perforated plate
- —— Eq. (31)

c) Relative Bubble Swarm Velocity

The gas hold-up is given by the ratio of the superficial gas velocity w_{SG} to the effective gas velocity w_G, i.e. the mean rising velocity of the bubbles. The latter is influenced by the properties of the phases, by the bubble size, the bubble density and the coalescence rate. At high bubble densities the velocity of the bubble swarm is low due to the mutual hinderance of the bubbles and the back-flow of the liquid. Coalescence results in growth of the bubbles and an increase in their rising velocity. This diminished the hold-up. Salt and/or alcohol additives stabilize the bubble size by formation of an ionic and/or a polar double layer at the interface which depresses the coalescence rate.

In Fig. 30 the relative gas velocities w_R are shown as a function of the superficial gas velocity w_{SG} for the perforated plate distributor, without and with additives. Without additives w_R passes a minimum at about $w_{SG} = 3.5$ cm/s. Up to this superficial gas velocity the bubbles rise in the bubble swarm independent of the bubbles in their neighbourhood. At higher gas velocities the bubble size increases by coalescence. Larger bubbles rise with the higher velocities. Therefore w_R increases after the minimum. One can recognize from Fig. 30 that in columns with additives and in the investigated range of w_{SG} that the bubble size does not change significantly, since w_R continuous to diminish with increasing w_{SG} as one would expect for a coalescence and turbulence free bubble swarm. According to Marucci [73]:

$$\frac{w_R}{w_T} = \frac{(1 - E_G)^2}{1 - E_G^{5/3}} \tag{32}$$

where w_T is the terminal bubble rising velocity for $w_{SG} = 0$, i.e. the rising velocity of a single bubble.

The fairly good agreement between the measured curves $w_R(w_{SG})$ and the calculated data (according to Marucci) indicates that the systems with additives are nearly free of coalescence.

Furthermore, one can recognize from Fig. 30 that for different additives with the perforated plate distributor the terminal bubble rising velocity w_T (i.e. the bubble size) is nearly constant. For porous plate distributors the terminal velocities w_T depend significantly on the type of additives (Fig. 31), i.e. the mean bubble size at the gas distributor is a function of the type of additives. One can also recognize from Fig. 31 that by means of salt additives. One can also recognize from Fig. 31 that by means of salt additives the position of the minimum in the $w_R(w_{SG})$ curve is shifted to higher superficial gas velocities, but in the investigated ranges of w_{SG} and of salt concentrations the minimum was not eliminated. Only with alcohol additives is the occurance of this minimum avoided. Unfortunately the agreement between the calculated and measured $w_R(w_{SG})$ curves is not as good as for the perforated plate distributor.

From Fig. 32 one can recognize that the concentration of the alcohols and their combination with salts influences the terminal velocities w_T as well as the slope of the curves $w_R(w_{SG})$. This again indicates the dependence of the coalescence rate on these factors.

The influence of the type of solute on w_T can be explained by the different surface tensions of these solutions. The surface tension in alcohol solutions decreases with the increasing hydrocarbon tail of the alcohols. w_T decreases in following sequence: methanol, ethanol, n-propanol due to the decreasing bubble diameter. Because of the depression of the surface tension the initial bubble size as well as the dynamical equilibrium diameter and/or $d_{B\,max}$ diminishes, the latter according to Eq. (18). On the other hand, salt of glucose additives increase the surface tension of pure water. Therefore in salt- or glucose solution one obtains slightly larger bubbles and higher terminal bubble velocities than in pure water.

The slope of the $w_R(w_{SG})$ curves is mostly influenced by the coalescence rate of the bubbles. Salt and/or alcohol-additives depress the coalescence rate due to the formation

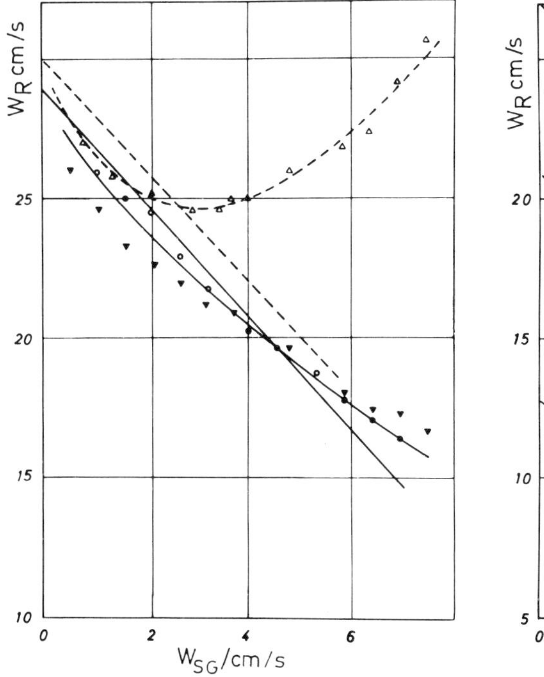

Fig. 30. Comparison of the measured relative velocities of the gas phase with regard to the liquid velocity w_R as a function of the superficial gas velocity with data calculated according to Marucci [73]. Perforated plate. w_{SL} = 2.2 cm/s.

△ H_2O

○ alcohols

▼ glucose

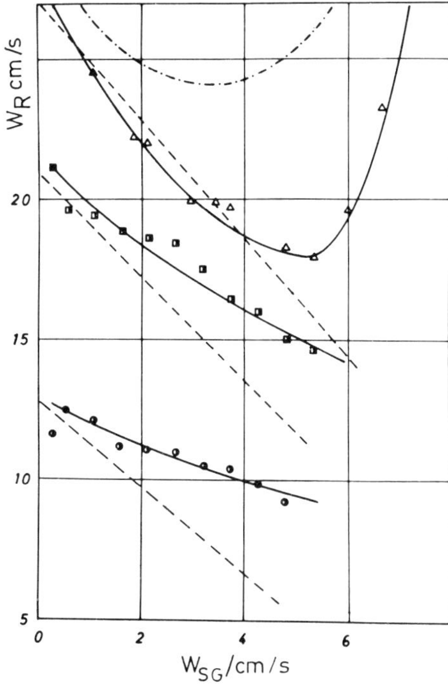

Fig. 31. Comparison of the measured relative velocities w_R as function of w_{SG} with data calculated according to Marucci [73]. Porous plate.

w_{SL} = 1.21 cm/s

—·—·— H_2O

△ NMP

◨ 1% CH_3OH

◑ 1% C_2H_5OH

– – – Marucci

of an electrical double-layer and/or a layer of oriented dipole molescule at the interface.

The small differences in the relative bubble swarm velocities for different concentrations of the same solute is mainly due to the differences in the bubble sizes. The large difference between the relative velocities w_R of bubbles in methanol and ethanol solutions (Fig. 31 and 32) is due to the different dynamical behaviour of these bubbles.

The behaviour of bubbles rising in pure liquids has been described by Hadamard [96] and Rybczynski [97], i.e. the interface is mobile and a circulation movement prevails which reduces the drag on the bubble in comparison with the one on the rigid sphere according to Stokes. If the liquid contains other constituents (alcohols, surface active agents) which have affinity for the interface, then the Hadamard-Rybczynski theory is not strictly applicable. Thus it is found that while large bubbles move in accord with this theory smaller bubbles move as if they were rigid bodies in accord with Stokes law.

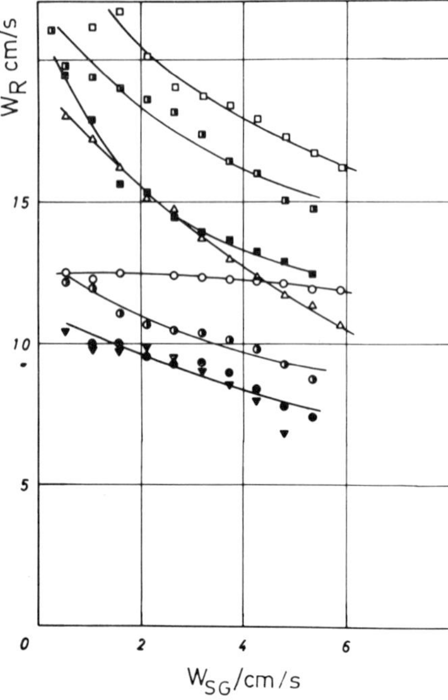

Fig. 32. Typical examples for w_R (w_{SG}) data with different compositions of the liquid. Porous plate w_{SL} = 1.21 cm/s.

□	0.5% CH_3OH	○	0.5% C_2H_5OH
◨	1% CH_3OH	◑	1% C_2H_5OH
□	2% CH_3OH	○	2% C_2H_5OH
△	1% CH_3OH	▼	1% C_2H_5OH/NMP

Thus for a clean bubble rising through a fluid containing a dissolved surface active substance, the substance will be adsorbed on the bubble surface and will be convected to the rear where it is accumulated and partly desorbed. A surface tension gradient will result which opposes the tangential shear stress and thereby stops the surface flow over the rear portion of the bubble.

As more material is adsorbed, a rigid cap grows as long as the surface tension at the rear continues to decrease, but when the latter reaches its minimum value any further accumulation of the substance cannot bring about a corresponding increase in the cap size since the existing surface tension gradient can not support the additional shear along the surface of the cap. Accordingly, if the bubble is small enough the cap will cover the entire surface owing to the effectiveness of the surface tension forces, while for a large bubble the surface film will collapse when the cap is very small. Therefore the transition of the interface from the freely circulating state to the more or less rigid state with decreasing bubble diameter (or increasing concentration of the surface active agent) is fairly sudden.

A comparison between the bubbles in 1% methanol and 1% ethanol solutions, both of them with the porous plate distributor and at superficial gas velocity w_{SG} = 3 cm/s (Figs. 31 and 32) by means of the models of Davis and Acrivos [94] as well as of Saville [95] shows, that in methanol solution the bubbles are nearly freely circulating (the drag coefficient is only 10% higher than the theoretical value for the freely circulating state and therefore the rising velocity is about 45% higher than the Stokes velocity) and that in contrast the bubbles in the ethanol solution are practically in the rigid state (the rising bubble velocity is only by 10% higher than the Stokes velocity). One can conclude that with the perforated plate and the alcohol/salt solutions freely circulating large bubbles are formed whose sizes remain constant. With the porous plate and the alcohol/salt solutions much smaller bubbles with "rigid caps" are formed. The size of these bubbles increases moderately along the column due to coalescence. In bubble columns with injector and ejector nozzles the Marucci equation (32) cannot be applied since near to the nozzle strong turbulence prevails and in this turbulence field bubbles coalescence and redispersion occurs due to the high energy dissipation density. However, since the terminal velocities $w_T \cong w_R(w_{SG})_{w_{SG}=0}$ one can draw qualitative conclusions with regard to the bubble size, and from the shape of the curve $w_R(w_{SG})$ one can assess the likelihood of coalescence. The curves $w_R(w_{SG})$ for water are quite similar in systems with different aerators (Figs. 30–34). In all of them the curves $w_R(w_{SG})$ pass through a minimum and increase again due to the strong coalescence which prevails in water. The terminal velocities w_T are very low with nozzles and in ethanol and/or sodium sulphate solutions (Figs. 33, 34). This indicates the presence of very small rigid

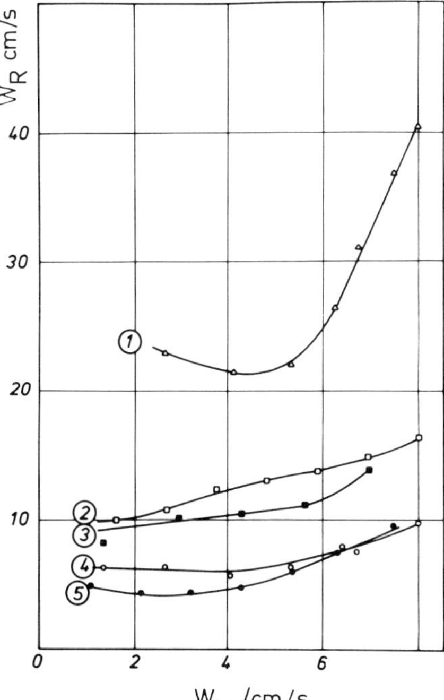

Fig. 33. Typical examples of $w_R(w_{SG})$ curves with different compositions of the liquids. Injector nozzle.

(1) H_2O w_{SL} = 1.2 cm/s
(2) 1% CH_3OH w_{SL} = 2.2 cm/s
(3) 1% CH_3OH/NMP w_{SL} = 2.2 cm/s
(4) 1% C_2H_5OH w_{SL} = 2.2 cm/s
(5) 10% Na_2SO_4 w_{SL} = 1.8 cm/s

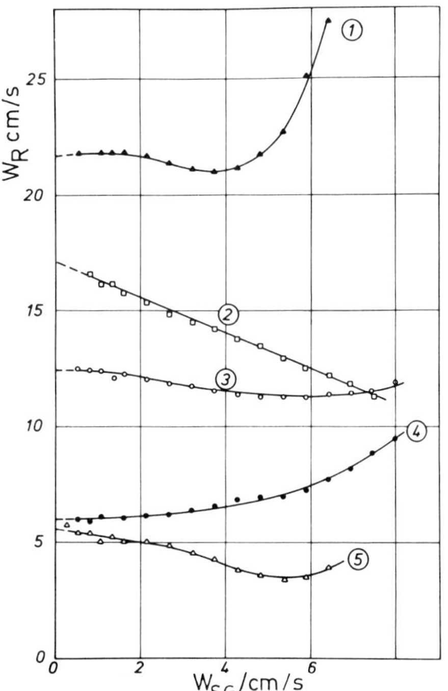

Fig. 34. Typical examples of $w_R(w_{SG})$ curves with different compositions of the liquids. Ejector nozzle.

(1) H_2O w_{SL} = 1.47 cm/s
(2) 1% CH_3OH w_{SL} = 1.5 cm/s
(3) 2% CH_3OH w_{SL} = 1.5 cm/s
(4) 0.5–2% C_2H_5OH w_{SL} = 1.52 cm/s
(5) 10% Na_2SO_4 w_{SL} = 1.66 cm/s

bubbles. w_T is slightly higher with the injector nozzle and moderately higher with the ejector nozzle for methanol than for ethanol solutions. This slight difference between ethanol and methanol systems can be explained by the same dynamical behaviour. The bubbles in both of these systems are rigid. The difference between the injector and ejector nozzles is due to the greater non-uniformity of the liquid velocity with the ejector nozzle which causes a higher coalescence rate in methanol solution too.

d) Bubble Size and Stability

The bubble size plays a decisive role with regard to the OTR. Because of the inherent stochastical character of the bubble column the bubble size is not uniform. The use of perforated plates leads to pressure fluctuations, porous plates involve a statistical pore size distribution, and with nozzles the stochastical character of the energy dissipation density due to turbulence causes a non-uniform size distribution of the initial bubbles. Coalescence and redistribution processes modify this initial distribution.
Typical bubble diameter distributions are shown in Figs. 35–41. In all systems with a high coalescence rate increasing the superficial gas velocity causes the bubble diameter distribution to be shifted to higher bubble diameters. This is also true of systems with a moderate coalescence rate (1% methanol-solution).

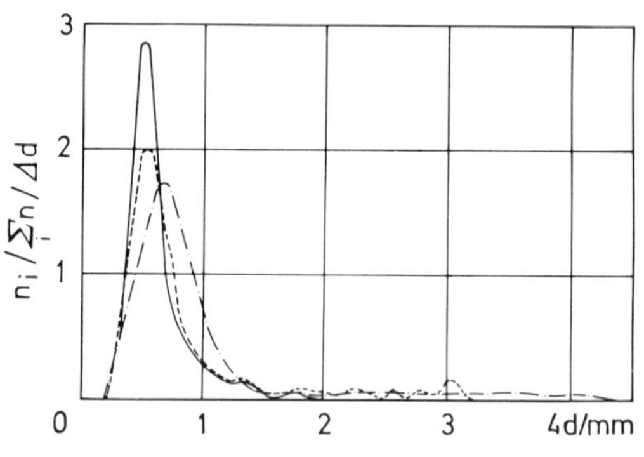

Fig. 35. Bubble size distribution in 1% CH_3OH solution. Injector nozzle. w_{SL} = 2.2 cm/s.
—— w_{SG} = 1.06 cm/s
– – – w_{SG} = 3.2 cm/s
–·–·– w_{SG} = 5.3 cm/s

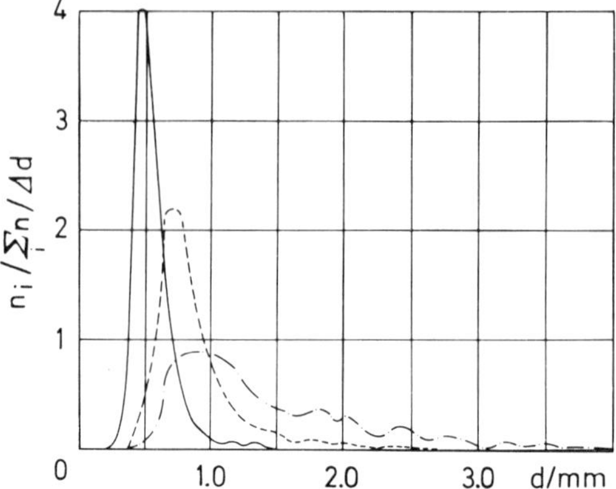

Fig. 36. Bubble size distribution in 1% CH_3OH solution. Ejector nozzle. w_{SL} = 1.5 cm/s.
—— w_{SG} = 1.06 cm/s
– – – w_{SG} = 3.2 cm/s
–·–·– w_{SG} = 5.3 cm/s

This shift to higher bubble diameters can clearly be recognized from Figs. 35 and 36. Again because of the higher bubble coalescence tendency with the ejector nozzle this shift is more significant than in systems with the injector nozzle. In systems with a very low coalescence rate (1% ethanol-, 10% sodium sulphate solution) increasing the superficial gas velocity results in the bubble diameter distribution being shifted to lower bubble diameters in such a way that the mean diameter only slightly changes since at higher gas superficial velocities the distributions have a longer tail (Fig. 37, 38, and 39). In systems with the lowest coalescence rate (10% Na_4SO_4 solution) this shift to smaller bubble diameters is the largest (Fig. 39). One finds this phenomenon in non-coalescing systems with the porous plate distributor (Fig. 40). The 1% Methanol + 1% salt-system behaves with regard to this shift as a non-coalescing system (Fig. 41). At very low gas flow rates the initial bubble diameter is independent of the gas flow rate but

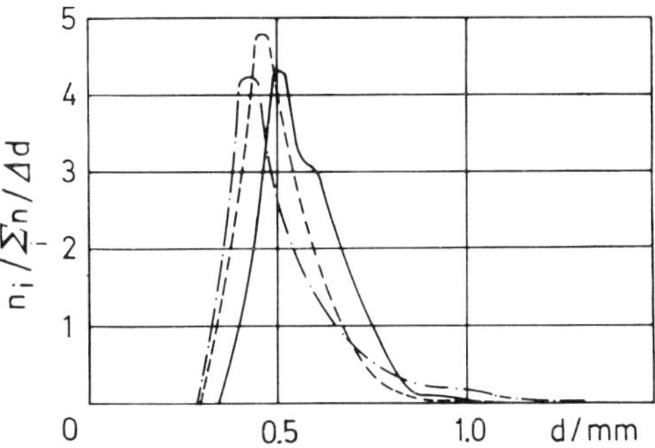

Fig. 37. Bubble size distribution in 1% C_2H_5OH solution. Injector nozzle.
w_{SL} = 2.2 cm/s.
—— w_{SG} = 1.06 cm/s
– – – w_{SG} = 3.2 cm/s
–·–·– w_{SG} = 5.3 cm/s

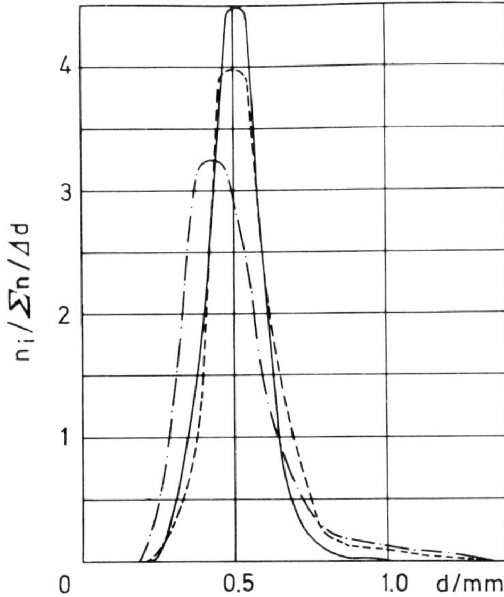

Fig. 38. Bubble size distribution in 1% C_2H_5OH solution. Ejector nozzle.
w_{SL} = 1.5 cm/s.
—— w_{SG} = 1.06 cm/s
– – – w_{SG} = 3.2 cm/s
–·–·– w_{SG} = 5.3 cm/s

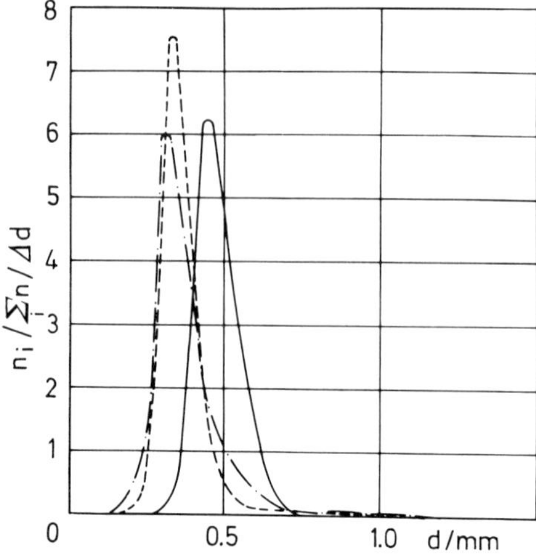

Fig. 39. Bubble size distribution in 10% Na_2SO_4 solution. Injector nozzle. w_{SL} = 2.2 cm/s.
—— w_{SG} = 1.06 cm/s
– – – w_{SG} = 3.2 cm/s
–·–·– w_{SG} = 5.3 cm/s

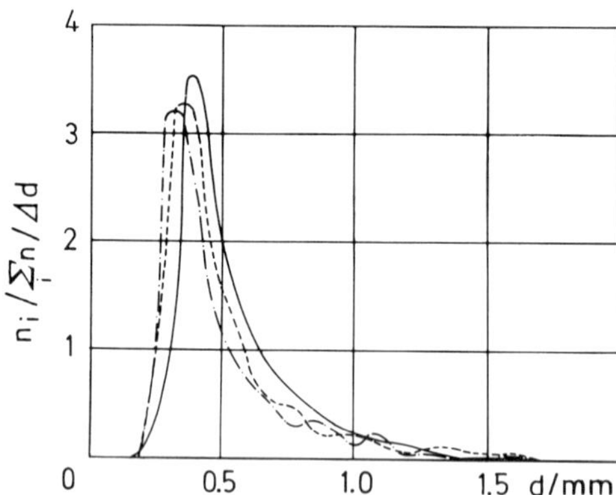

Fig. 40. Bubble size distribution in 10% Na_2SO_4 solution. Porous plate. w_{SL} = 2.2 cm/s.
—— w_{SG} = 1.06 cm/s
– – – w_{SG} = 4.0 cm/s
–·–·– w_{SG} = 6.3 cm/s

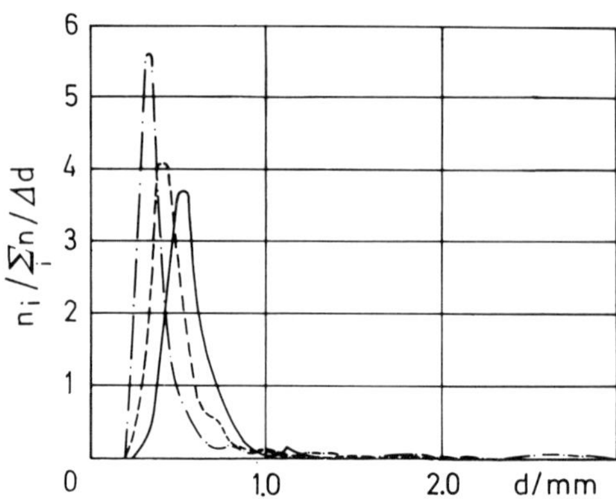

Fig. 41. Bubble size distribution in 1% CH_3OH + NMP solution. Injector nozzle. w_{SL} = 2.2 cm/s.
—— w_{SG} = 1.06 cm/s
– – – w_{SG} = 3.7 cm/s
–·–·– w_{SG} = 5.3 cm/s

at somewhat higher rates the initial bubble diameter increases with increased gas flow rate [44]. Therefore this shift to smaller bubble diameters with the porous plate can not be caused by a change in the initial bubble size, if one assumes the validity of the Kumar-Kuloor-Model for low gas flow rates. However, Marmur and Rubin [49] have shown that for orifices at high gas flow rates the inertia forces play an important role. The forming bubble tends to detach from the orifice earlier due to rotational motion induced in the liquid. Similar effects can also appear with the porous plate. However, at the gas flow rates used in this investigation the shift is probably caused by another effect. At medium gas flow rates the Rayleigh relations (10) apply [110, 111] and these demand an independence of the initial bubble size on the gas flow rate. At high gas flow rates the bubble diameter is controlled by the dynamic pressure of the local turbulence at the aerator. With increased gas flow rate the intensity of the local turbulence increases, hence the dynamical equilibrium bubble size diminishes. Therefore, one can assume, that the local turbulence at the aerator dominates, if with increasing gas flow rate a reduction of the size of bubbles occurs in systems with strong hinderance of the coalescence. Because of the extremely complex processes which occur during the gas dispersion in the liquid phase by means of ejector and injector nozzles, an explanation of this shift of the distribution to lower bubble diameters for nozzle aerators is very difficult. Due to the coalescence hinderance in cultivation media the actual bubble diameter does not necessarily correspond to the dynamical equilibrium diameter controlled by the prevailing energy dissipation density. The actual bubble diameter will lag behind the dynamic equilibrium diameter, if the latter increases along the column due to the decreasing rate of energy dissipation.

In the momentum exchange tube the rate of energy dissipation is the highest at the plane normal to the axis, where the liquid jet bounces against the tube wall and the gas is dispersed. Downstream from this plane the rate of energy dissipation decreases. It is quite probable that as long as the bubbles are in the momentum exchange tube they correspond to the dynamical equilibrium bubbles. At the exit of this tube the rate of energy dissipation quickly decreases (with the fourth power of the distance). It is assumed that downstream of the momentum exchange tube the bubble size is nearly preserved, i.e. the coalescence is "frozen". Since at higher gas flow rates the dispersion of the gas phase occurs in the momentum exchange tube nearer to the tube exit [76], the "frozen" bubble diameter will be smaller than the one for lower gas flow rates. However, no experimental facts are available yet to support this assumption. Because the bubble size distributions are not symmetrical the statistical mean and the "Sauter" mean diameters are different. This difference is especially large in systems with significant coalescence rates. Since the large bubbles have a larger weight than the small ones in the "Sauter" mean, the "Sauter" mean is higher for coalescence-systems than the statistical mean. A comparison between the "Sauter" and statistical mean bubble diameters shows (Fig. 42 and 43) that in systems with a low coalescence rate (1% ethanol, 10% sodium sulphate) the difference is low and with increasing coalescence rate (Methanol + salt → methanol) the difference increases. Since the "Sauter" mean diameters are used to calculate the specific interfacial areas, they are considered in the further discussion.

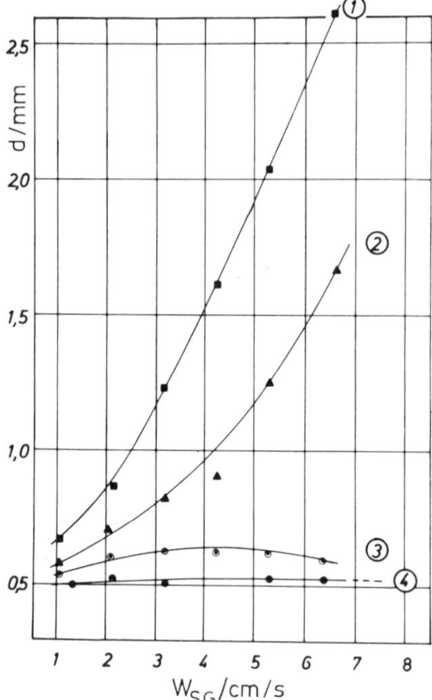

Fig. 42. "Sauter" mean diameter of the bubbles as function of the superficial gas velocity. Ejector nozzle.
(1) 1% CH_3OH
(2) 1% CH_3OH + 1% salt
(3) 1% C_2H_5OH
(4) 10% Na_2SO_4

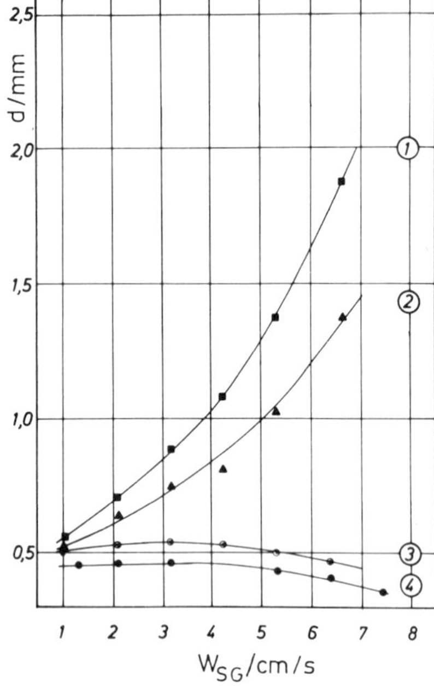

Fig. 43. Statistical mean diameter of the bubbles as function of the superficial gas velocity. Ejector nozzle.
(1) 1% CH_3OH
(2) 1% CH_3OH + 1% salt
(3) 1% C_2H_5OH
(4) 10% Na_2SO_4

In Fig. 44 the Sauter bubble diameters are shown as a function of the superficial gas velocity for the perforated and porous plate distributors and for liquids of various composition. One can recognize that bubbles with the largest size are produced in water with the perforated plate (1). The addition of salt (1% solution) does not change the Sauter mean diameter. By applying alcohols (1% CH_3OH or 1% C_2H_5OH) the diameter is moderately diminished (2). The porous plate produces bubbles in water (3) which are smaller than those produced by the perforated plate. The influence of additives on the bubbles diameter when using the porous plate is much larger than in systems with the perforated plate. The diameter decreases in the following sequence: 1% salt solution (4) > 0.5% methanol solution (5) > 1% ethanol solution (6). The influence of the composition of the liquid on the Sauter diameter is shown in Fig. 42 for the ejector nozzle and in Fig. 45 for the injector nozzle. In non-coalescencing systems (1% ethanol, 10% sulphate solution) the influence of the superficial gas velocity on the Sauter diameter is relatively slight, especially with the ejector nozzle. With increasing tendency for coales-

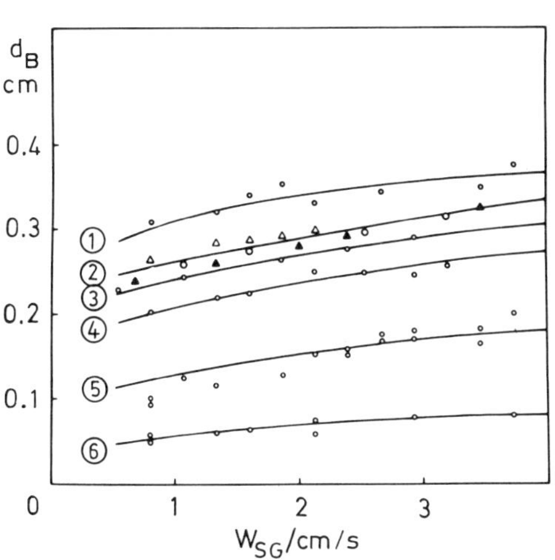

Fig. 44. "Sauter" mean diameter of bubbles as function of the superficial gas velocity.

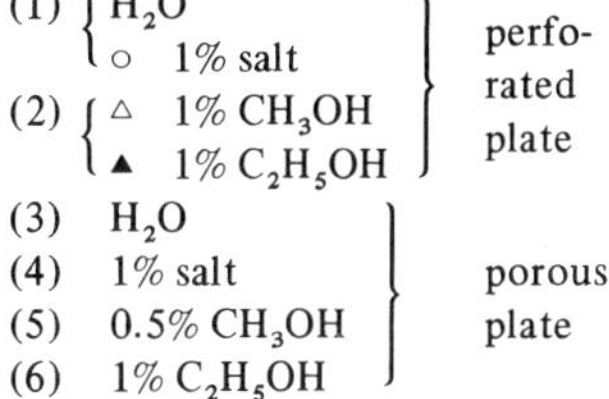

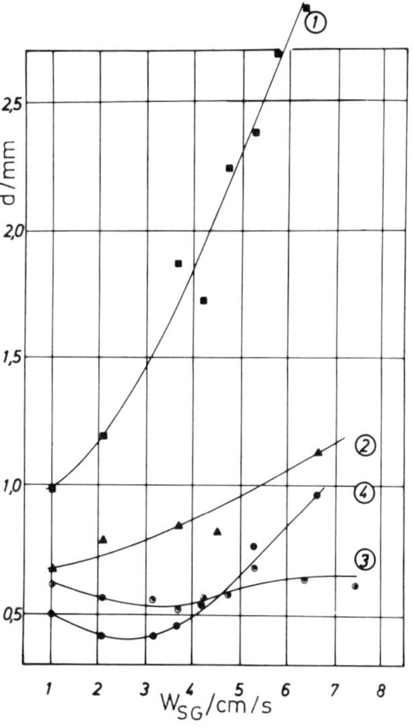

Fig. 45. "Sauter" mean diameter of bubbles as function of the superficial gas velocity. Injector nozzle.

(1) 1% CH_3OH
(2) 1% CH_3OH + 1% salt
(3) 1% C_2H_5OH
(4) 10% Na_2SO_4

cence the influence of the gas velocity on the bubble diameter also increases. The influence of the aerator type on the bubble diameter is given for 1% methanol solution (Fig. 48) and 10% sodium sulphate solution (Fig. 49). In general the bubble diameter decreases in the following sequence: perforated plate > porous plate > injector nozzle > ejector nozzle (Figs. 44 and 46–49).

To explain the behavior of the bubbles in different systems their stability has to be considered. According to Berghmans [38] the stability of bubbles can be estimated when the bubbles are moving upwards in liquids of small viscosity because of their buoyancy, if one considers the buoyancy interfacial and gravitational forces and neglects the viscous and inertia forces. Some typical examples are shown in Fig. 50. In demineralized water the bubbles are always on the boundary between the stable and unstable regions independently on the aerator type. This indicates that the coalescence rate is high and at each position in the column the local dynamical equilibrium bubble diameter prevails.

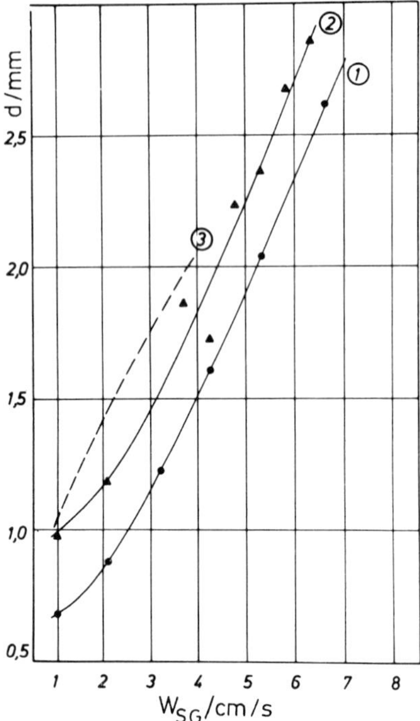

Fig. 46. "Sauter" mean diameter of bubbles as function of the superficial gas velocity. 1% methanol solution. Comparison of different aerator types.
(1) ejector nozzle
(2) injector nozzle
(3) porous plate

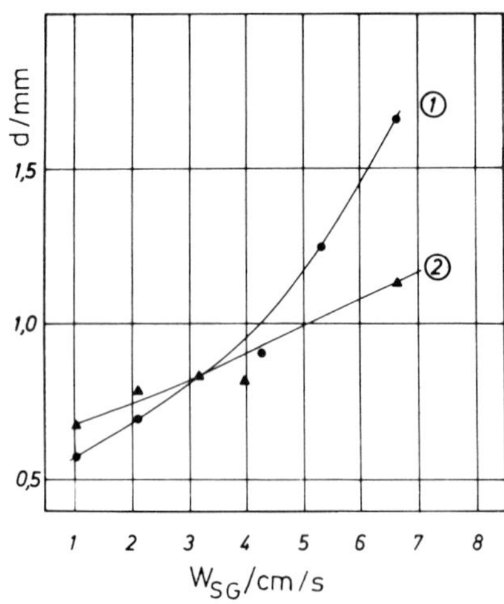

Fig. 47. "Sauter" mean diameter of bubbles as function of the superficial gas velocity. 1% methanol–1% salt solution. Comparison of the nozzle aerators.
(1) ejector nozzle
(2) injector nozzle

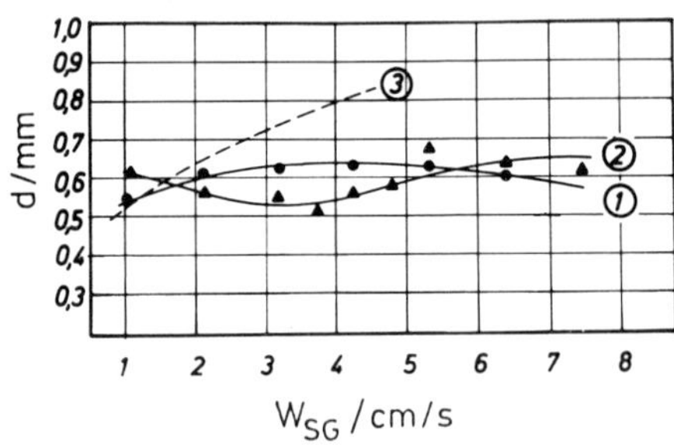

Fig. 48. "Sauter" mean diameter of bubbles as function of the superficial gas velocity. 1% ethanol solution. Comparison of different aerator types.
(1) ejector nozzle
(2) injector nozzle
(3) porous plate

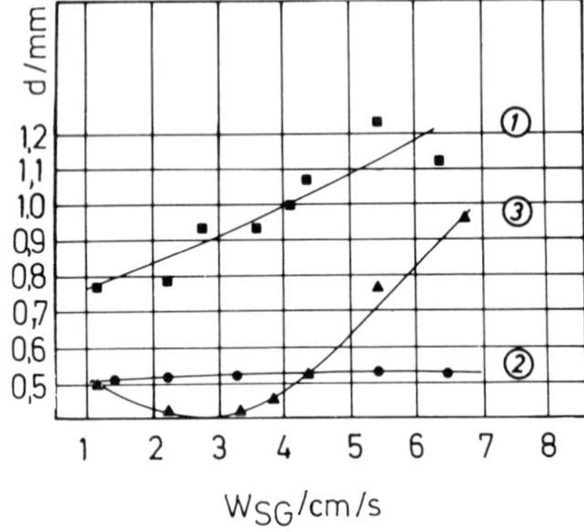

Fig. 49. "Sauter" mean diameter of bubbles as function of the superficial gas velocity. 10% Na_2SO_4 solution. Comparison of different aerator types.
(1) porous plate
(2) ejector nozzle
(3) injector nozzle

Fig. 50. Bubble stability diagram:
—— boundary between stable and unstable regions [9]
X results of Haberman-Morton [10]
results of the authors:
□ porous plate
▲ perforated plate
○ injector nozzle

w_{SG} cm/s	2	6
1% CH_3OH	(1)	(5)
1% CH_3OH + NMP	(2)	(6)
1% C_2H_5OH	(3)	(7)
10% Na_2SO_4	(4)	(8)
H_2O	(9)	(10)
w_{SG} cm/s	4	
0.5% CH_3OH + NMP	(11)	
0.5% C_2H_5OH + NMP	(12)	

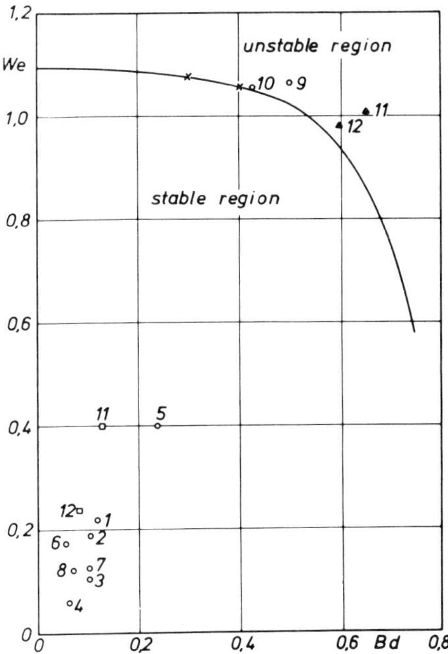

With the perforated plate the bubbles are also on (or near to) the boundary between the stable and unstable regions, independent of the composition of the liquid. This indicates that in these systems the dynamic equilibrium bubble diameter also dominates. To investigate whether the dynamical equilibrium diameter is attained by coalescence of initially smaller bubbles or whether they come into equilibrium by redistribution of initially larger bubbles, the initial bubble diameters were calculated for the perforated plate by means of Eq. (8) of Kumar and Kuloor [44] and Eq. (10) of Rayleigh [110, 111].

In water with the perforated plate the initial bubble diameter calculated by Eq. (10) is always smaller than the measured diameter. Equation (8) also yields a smaller initial bubble size than measured for $w_{SG} < 6$ cm/s. Since it is improbable that Eq. (8) is also valid for $w_{SG} > 6$ cm/s ($(w_{SG})_{cr} = 3.4$/s) the size of bubbles measured in the column must arise from coalescence. In alcohol/salt solutions with the perforated plate the calculated initial bubble diameters are also less than the measured diameters. The latter are equal to the dynamical equilibrium bubble diameters. Therefore in cultivation media the bubble diameter is always determined by the dynamical equilibrium diameter, if a perforated plate with hole diameter $\geqslant 0.5$ is used.

Because of the extremely high rate of local energy dissipation which prevails near the two-component nozzles, the initial bubble diameter is very small (e.g. $d_s \cong 36\ \mu m$ at $\frac{E}{V} = 300\ kW/m^3$) [74]. Since the dispersion of the gas phase occurs in the "momentum exchange tube" of the nozzle, the high rate of energy dissipation is maintained in the two-phase flow until it leaves the tube and expands into the bubble column. As a result of the expansion the local mean flow velocity of the liquid w_L^* decreases with increasing distance Z from the outlet of the tube according to $w_L^* \sim Z^{-2}$ and the scale of turbulence l increases according to $l \sim Z$ [91]. Since energy is dissipated most effectively by small scale turbulence [91], the increase of the scale of turbulence as well as the decrease of the intensity of turbulence due to the decrease of w_L^* diminish the turbulent shear stress τ according to $\tau \sim w_L^{*2}$, i.e. $\tau \sim Z^{-4}$ [91].

The rate of energy dissipation decreases and the dynamical equilibrium bubble size increases quickly in the expanding bubble flow until it reaches the column wall. In the core region of the fully developed bubble flow (i.e. with exception of the entrance region) nearly all of the properties of the two phase flow (mean local liquid velocity, mean local bubble velocity, local bubble frequency, local gas hold-up, local intensity and relative local intensity of turbulence in the liquid, relative standard deviation of bubble velocity) are nearly independent of the column radius [92, 93].

Only for small column diameters does the increase in the gas hold-up near to the column wall have to be considered. For small column diameters and high liquid velocities (e.g. in the "momentum exchange tube" of two-phase nozzles) the steep increase of the shear stress near to the wall is also significant [92, 93]. However, bubble columns usually involve large diameters and low liquid velocities and consequently changes at the wall can be neglected. Therefore one can calculate the properties of the fully developed bubble flow at every cross section of the column.

The longitudinal change of the turbulent shear stress τ in the fully developed bubble flow along the column depends on the complex interchange of following processes:

1. Because of the pressure drop the gas hold-up increases and hence the effective cross section of the liquid flow decreases. This increases the linear liquid velocity and the intensity of the turbulence.
2. Energy is dissipated due to the friction of liquid and bubbles, and the bubble compression, oscillation, rotation etc.. This diminishes the intensity of the turbulence.
3. With increasing bubble size the scale of turbulence increases and the rate of energy dissipation decreases.

Usually a decrease in the turbulent shear stress τ and an increase of the dynamical equilibrium bubble diameter occurs in the fully developed bubble flow along the column. The extent of these changes are much less than the corresponding changes in the expanding region.

Since the coalescence rate in water is very high, the local dynamical equilibrium bubble diameter controls the bubble size at every point in the column. If two-component nozzle aerators are used with water the initial bubble diameter quickly increases in the expanding bubble flow and then grows slowly by coalescence in the fully developed bubble flow.

In alcohol/salt solutions and with two-component nozzle aerators the extremely small initial bubble size is also enlarged by coalescence. However, the coalescence stops in a region where the energy dissipation density is relatively high. The local dynamical equilibrium diameter of the bubble in the column can not be reached. Therefore all of these bubbles are in the stable range and it follows that the bubble diameter depends on the aerator type (Fig. 50).

The effective linear gas velocity in the bore holes of the plate aerators decreases with

increasing free surface area. With the reduction of the hole diameter and the local gas velocity the initial bubble size in the bubbling gas range also decreases [44, 75]. Therefore the initial bubble diameter must be smaller with the porous plate (mean porosity 40%, pore diameter $< 20\ \mu m$) compared with the perforated plate (free surface area 0.23%, hole diameter 0.5 mm) used in these investigations. The same is valid in the gas jet range, since the initial bubble diameter is reduced with diminishing hole diameter. In the bubbling gas range as well as in the gas jet range the calculated initial bubble sizes are smaller for porous plate than the measured ones. Hence the bubbles must suffer coalescence and grow along the column depending on their tendency for coalescence. Because of the slight difference between the initial and the dynamic equilibrium bubble size for the perforated plate the bubbles quickly attain the latter and hence the bubble size is controlled by the dynamical equilibrium between the dynamic pressure force and surface tension force. In alcohol/salt solutions the coalescence is hindered. Therefore in alcohol/salt mediums with the porous plate, when the initial bubble size is much smaller than the dynamical equilibrium bubble size, the latter is not attained. Therefore the bubbles are in the stable range as can be seen from the stability diagram (Fig. 50).
Kozo Koide [65] recommended an empirical correlation for the bubble diameter in coalescencing liquids (water) and non-coalescencing liquids (alcohol/salt solution) with porous plate distributors:

$$d_B \left(\frac{g \rho_L}{\delta_p \sigma}\right)^{1/3} = A \left(\frac{Fr^*}{We^{*0.5}}\right)^a \qquad (33)$$

where $a = 0.10$, $A = 0.64$ for non-coalescencing liquids

$a = 0.16$, $A = 1.65$ for coalescencing liquids

$We^* = \dfrac{w_{SG}^2 \delta_p \rho_L}{E_p^2 \sigma}$ modified Weber number

$Fr^* = \dfrac{w_{SG}^2}{E_p^2 \delta_p}$ modified Froude number

δ_p pore diameter and

E_p porosity of the porous plate.

The agreement between the measured and the calculated diameters [Eq. (33)] are good for water and methanol solutions. However, for ethanol solutions the calculated diameter is nearly twice as high as the measured value. The difference between the methanol and ethanol systems obviously can not be explained by the different surface tension alone. More experimental results and better theory are necessary to calculate the bubble diameter in liquid mixtures with porous plate distributors.

e) Specific Interfacial Area

The interfacial area a' can be considered with regard to the volume of the bubble layer V or to the volume of the liquid V_L.

i.e.

$$A = \frac{a'}{V} = \frac{6\,E_G}{d_S} \tag{34}$$

and

$$a = \frac{a'}{V} \cdot \frac{1}{E_L} = \frac{a'}{V_L} \tag{35}$$

where $E_L = (1 - E_G)$.

In this paper the specific interfacial area *'a'* defined by Eq. (34) is used mainly to compare the results with the data of other groups of workers. The relative mean gas hold-up E_G and the "Sauter" mean diameter d_s, which appear in Eq. (34) have already been discussed.

In Fig. 51 the specific interfacial area a'/V_L is plotted as a function of the superficial gas velocity w_{SG} for the perforated plate with different liquids. One can recognize from Fig. 51 that the increase of a'/V_L due to additives when using the perforated plate is non-specific, i.e. similar to the increase in the gas hold-up. When using either the porous

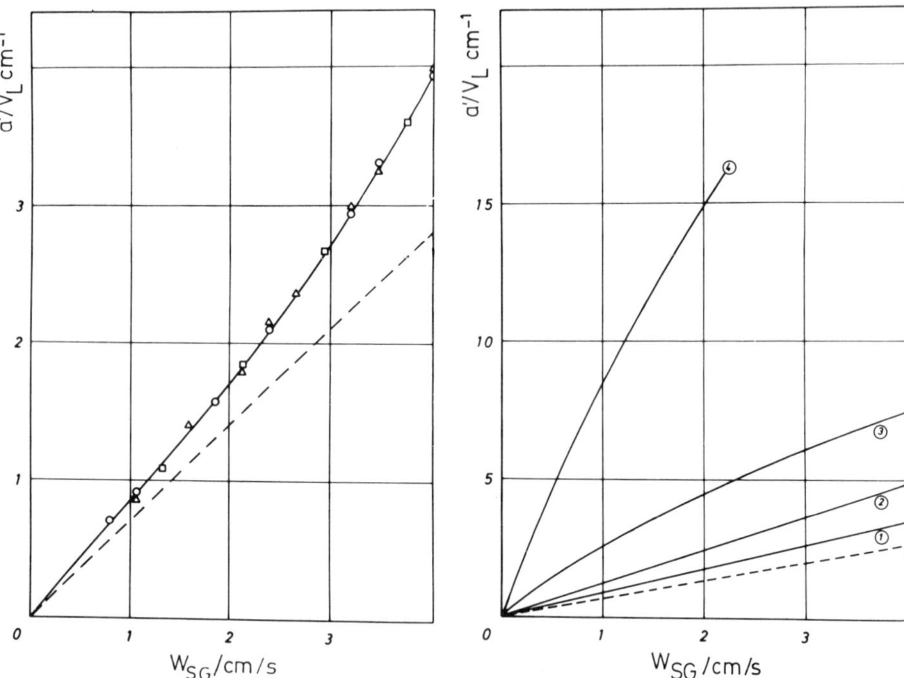

Fig. 51. Typical specific interfacial areas a'/V_L as function of the superficial gas velocity w_{SG}.
Perforated plate.

□ 0.5% CH_3OH

○ 1% C_2H_5OH

△ 1% salt

--- H_2O

Fig. 52. Typical specific interfacial areas a'/V_L as function of the superficial gas velocity w_{SG}.
One-stage systems.

Porous plate	Perforated plate
(1) H_2O	alcohol + salt (identical with curve (2))
(2) 1% salt	--- H_2O
(3) 0.5% CH_3OH	
(4) 1% C_2H_5OH	

plate (Fig. 52), the injector nozzle (Fig. 53), or the ejector nozzle (Fig. 54) aerators the composition of the liquid influenced a'/V_L and/or a'/V. The specific interfacial area increases in following sequence:
H_2O < 1% CH_3CH solution < 1% CH_3OH – 1% salt-solution < 1% C_2H_5OH – 1% salt-solution < 1% C_2H_5OH solution < 10% Na_2SO_4 solution (with ejector nozzle for w_{SG} < 4 cm).
Furthermore, the influence on a'/V of the aerator type is large as can be seen for 1% methanol solution (Fig. 55), 1% methanol – 1% salt-solution (Fig. 56), 1% ethanol solution (Fig. 57); 1% ethanol – 1% salt solution (Fig. 58) and 10% sodium sulphate solution (Fig. 59).
In ethanol-salt solution with the ejector nozzle a few large bubbles are present, hence the Sauter diameter increases and the specific interfacial area decreases with increasing superficial gas velocity for w_{SG} > 4 cm/s. By disregarding these few large bubbles the same a'/V-values are found for both the injector and the ejector nozzles.

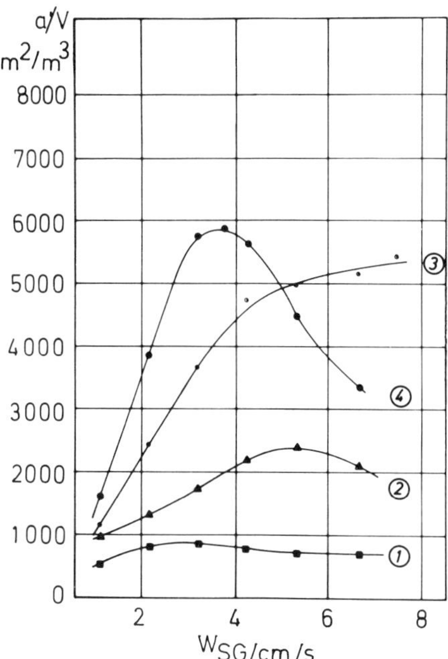

Fig. 53. Specific interfacial area a'/V as function of the superficial gas velocity. Injector nozzle. Influence of the composition of the liquid.
(1) 1% CH_3OH
(2) 1% CH_3OH–1% salt
(3) 1% C_2H_5OH
(4) 10% Na_2SO_4

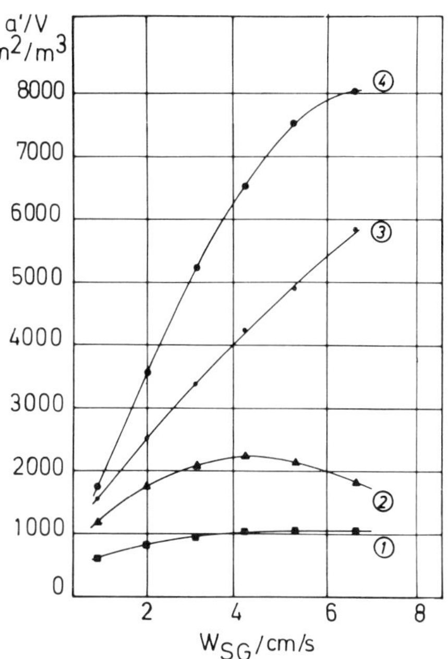

Fig. 54. Specific interfacial area a'/V as function of the superficial gas velocity. Ejector nozzle. Influence of the composition of the liquid.
(1) 1% CH_3OH
(2) 1% CH_3OH–1% salt
(3) 1% C_2H_5)H
(4) 10% Na_2SO_4

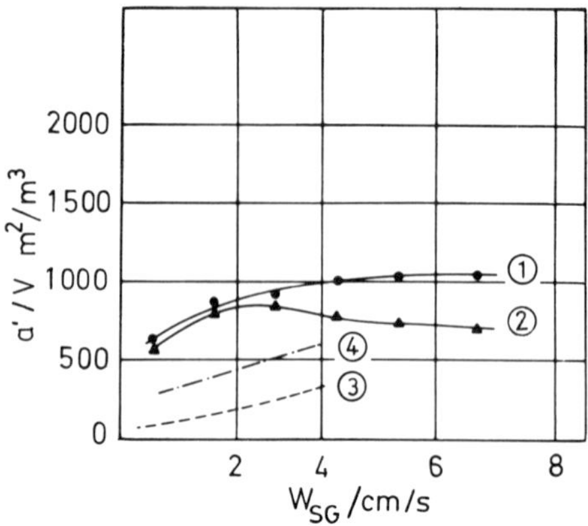

Fig. 55. Specific interfacial area a'/V as function of the superficial gas velocity. 1% methanol solution. Comparison of different aerator types.
(1) ejector nozzle
(2) injector nozzle
(3) perforated plate
(4) porous plate

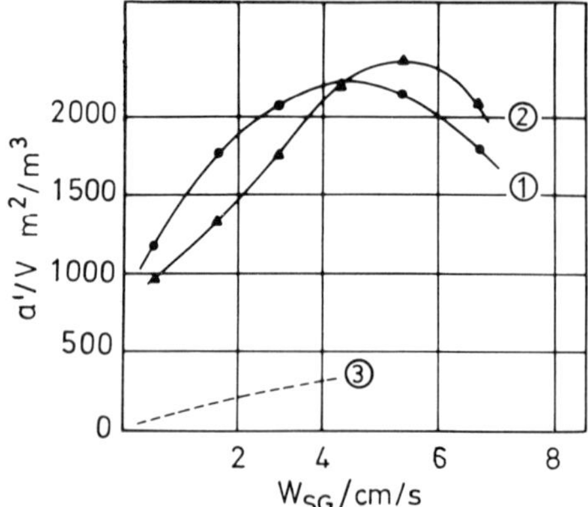

Fig. 56. Specific interfacial area a'/V as function of the superficial gas velocities. 1% methanol–1% salt solution. Comparison of different aerator types.
(1) ejector nozzle
(2) injector nozzle
(3) perforated plate

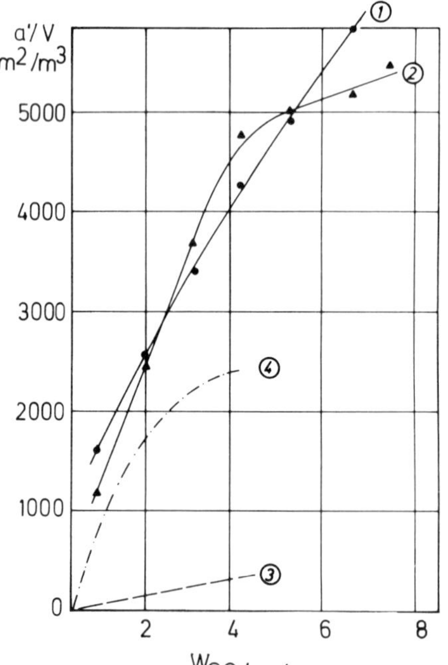

Fig. 57. Specific interfacial area a'/V as function of the superficial gas velocity. 1% ethanol solution. Comparison of different aerator types.
(1) ejector nozzle
(2) injector nozzle
(3) perforated plate
(4) porous plate

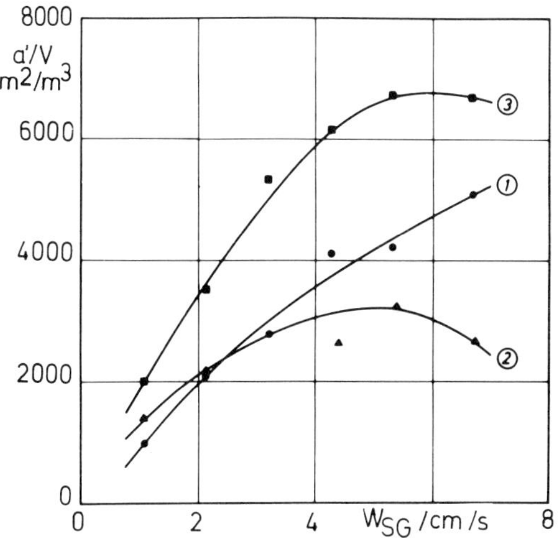

Fig. 58. Specific interfacial area a'/V as function of the superficial fas velocity. 1% ethanol–1% salt solution. Comparison of different aerator types.
(1) porous plate
(2) ejector nozzle
(3) injector nozzle

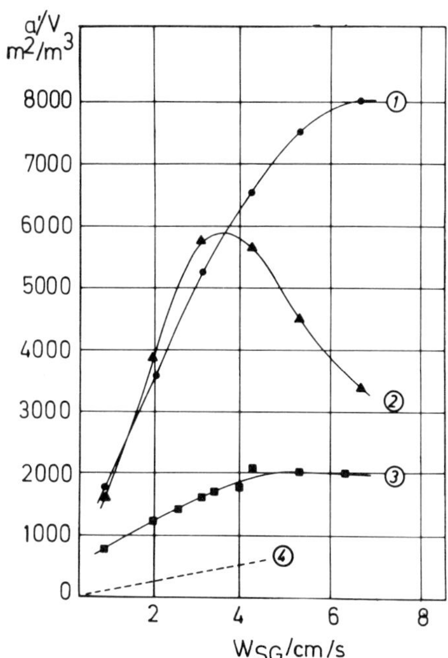

Fig. 59. Specific interfacial area a'/V as function of the superficial gas velocity. 10% sodium sulphate solution. Comparison of different aerator types.
(1) ejector nozzle
(2) injector nozzle
(3) porous plate
(4) perforated plate

The specific interfacial area increases in following sequence:
perforated plate < porous plate < injector nozzle, ejector nozzle (Figs. 55–59).
High specific interfacial areas can be produced by nozzles in bubble columns for the fermentation media of SCP production, e.g.

$a'/V > 2000\ m^{-1}$ for methanol medium
$a'/V > 5000\ m^{-1}$ for ethanol medium.

The energy required to produce these areas is discussed below.

f) Energy Requirement

The rate of the dissipated energy per unit volume is calculated for the perforated plate from the work which is needed to overcome the static liquid pressure by the gas [77].

$$E = -\phi_{m,G}\, \Delta\, [\frac{RT}{M} \ln p + \frac{1}{2}\, w_G^2 + gx] \tag{36}$$

where Δ = the change of the terms in the parenthesis along the height of the bubble column
$\phi_{m,G}$ = mass flow of the gas
p = local pressure.

The energy required for the formation of the bubbles at the gas distributor is not considered in Eq. (36) because it is sufficiently small to be neglected, as shown below: According to Burkel [72] the pressure needed for the separation of a bubble from the gas distributor is given by

$$P = \frac{4\,\sigma}{D_{ö},\,\delta_p} \tag{37}$$

where $D_{ö}$ = hole diameter of perforated plate,
δ_p = pore diameter of porous plate.

The separation pressures for water resulting from Eq. (37) are

$P = 5.80 \cdot 10^{-3}$ atm (perforated plate)

and

$P = 16.78 \cdot 10^{-3}$ atm (porous plate)

both of which are negligible.

To avoid error in the evaluation of the energy requirement for the porous plate and nozzle aerators the measured pressure drop across the aerator-bubble-column system was used. In these calculations the energy losses of the compressors were not considered.

In Fig. 60 the specific interfacial area is plotted as a function of the rate of energy dissipation E/V for some typical perforated and/or porous plate systems both from the present work and from Reith [77].

Fig. 60 indicates that the specific interfacial areas obtained in the present work for both of the gas distributors are higher than the corresponding values of Reith [77] [(4) and (5) in Fig. 60].

Curve 6 of Reith in Fig. 60 was measured in a bubble column (diameter: 14 cm, height: 200 cm and 340 cm) similar to the column used in the present work. However, the perforated plate distributor of Reith had holes of 2 mm diameter. The interfacial areas presented in this paper are significantly larger than the corresponding values of Reith, especially for the porous plate with ethanol as the solute, since the quality of gas dispersion was better.

In Fig. 61 various apparatus configurations containing sulphite oxidation systems are compared with regard to the specific interfacial area a'/V. One can recognize that in bubble columns the choice of a suitable aerator type plays an important role [compare (2), (3), and (6) in Fig. 61]. The bubble column with a porous plate (2) is a very economical fermentor, because high specific interfacial areas can be produced by very low energy inputs. Bubble columns with ejector and/or injector nozzle (6) are also very effective fermentors, because very high specific interfacial areas can be producted with medium energy inputs.

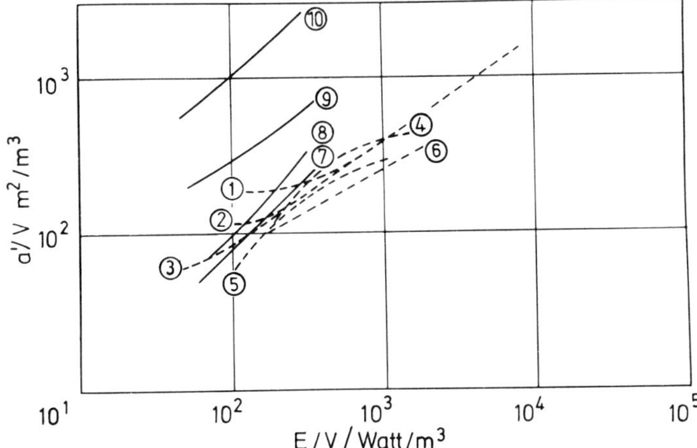

Fig. 60. Comparison of the specific interfacial areas a'/V as function of the rate of energy dissipation E/V for some typical one-stage systems of the present work with data of Reith [77].

(1) H_2O (stirred tank)
(2) H_2O (stirred tank)
(3) H_2O (stirred tank)
(4) H_2O (bubble column)
(5) H_2O (bubble column)
(6) H_2O (bubble column)
} Reith

(7) perforated plate H_2O
(8) perforated plate CH_3OH, C_2H_5OH, salts
(8) porous plate H_2O
(9) porous plate 0.5% CH_3OH
(10) porous plate 1% C_2H_5OH
} present work

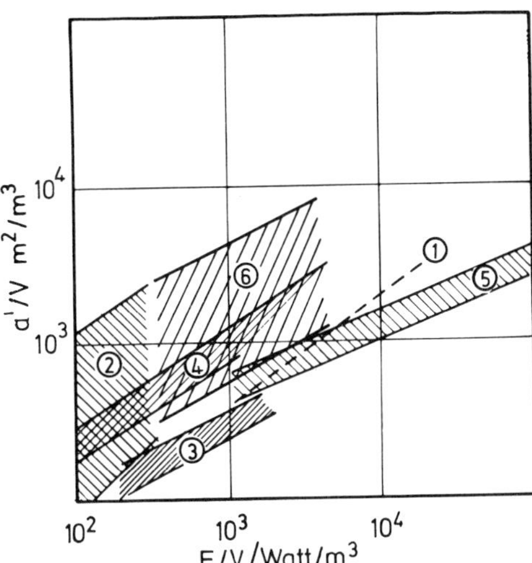

Fig. 61. Comparison of specific interfacial areas a'/V as function of the rate of energy dissipation E/V in sulphite oxidation systems.

(1) stirred tank reactor (Reith [77])
(2) bubble column-porous plate (authors)
(3) bubble column perforated plate (Reith [77])
(4) ejector nozzle (Nagel et al. [42])
(5) packed bed column (Nagel et al. [42])
(6) bubble column–nozzle injector (authors)

g) Mass Transfer Coefficients

Since the measurement of the volumetric mass transfer coefficient ($k_L a$) is more accurate than that of k_L the volumetric mass transfer coefficients are mainly considered. The estimation of the volumetric mass transfer coefficients was carried out by the stationary technique: the longitudinal concentration profile of the oxygen in the culture medium was measured by means of oxygen electrodes at up to 30 points along the column. A dispersion model was used to evaluate the model parameters. The volumetric mass transfer coefficients ($k_L a$) and the *Bo*-numbers of the longitudinal dispersion ($Bo = \frac{w_L H}{D_L B}$) in the liquid phase were evaluated by fitting the calculated longitudinal concentration profiles of oxygen in the liquid to be measured profiles. The stationary method used in the present work yields accurate ($k_L a$) values. As can be seen from Fig. 3 the shape of the longitudinal concentration profiles is not very sensitive to the *Bo*-numbers. Therefore the *Bo*-numbers, estimated by this procedure, are not very accurate. Therefore the estimation of the *Bo*-numbers was carried out by separate measurements by means of a tracer technique. Since with nozzles the saturation of the liquid by oxygen occurred very quickly in the column, only a few points of the longitudinal concentration profiles could be used for curve fitting. It follows that the volumetric mass transfer coefficients, measured with nozzles, are less accurate than the ones evaluated for porous and perforated plates; only the latter are discussed in detail.

In the investigated range of the superficial gas velocities (w_{SG} = 0.3 to 4 cm/s) the volumetric mass transfer coefficients ($k_L a$) increase with w_{SG} and are proportional to it.

In pure water ($k_L a$) is higher for the porous plate than for the perforated plate and it is independent of w_{SL} (Fig. 62). The latter is in agreement with the result of Chang [78], who measured ($k_L a$) in bubble columns with tap water.

In bubble columns with the perforated plate distributor the salt and alcohol additives caused a non-specific increase of ($k_L a$) analogous to E_G (Fig. 63), while with glucose additives the ($k_L a$) is greater than with alcohols and salts (Fig. 64). The two types of plate distributor produced different large mass transfer coefficients with both pure water and salt solutions.

In systems with the porous plate the increase of ($k_L a$) is specific to the type of the additives and partly to the concentrations (Fig. 65). The combination of porous plate with ethanol can produce ($k_L a$) values which are ten times as high as those with the perforated plate (Fig. 65). The effect of the concentration of the additives on ($k_L a$) is slight for methanol-salt-systems as well as for ethanol-salt-systems in contrast to their effect on E_G. The variance of the measured values depends on the distributor type and additives: the mean relative deviations from the regressive equation are 11% for perforated plate with additives, and for the porous plate depend on the solute, i.e.

H_2O	9%
H_2O/salts	8.5%
0.5% CH_3OH	10.2%
1% and/or 2% CH_3OH	18%
CH_3OH + salts	14.4%
0.5 to 2% C_2H_5OH	23.3%
C_2H_5OH + salts	20.5%

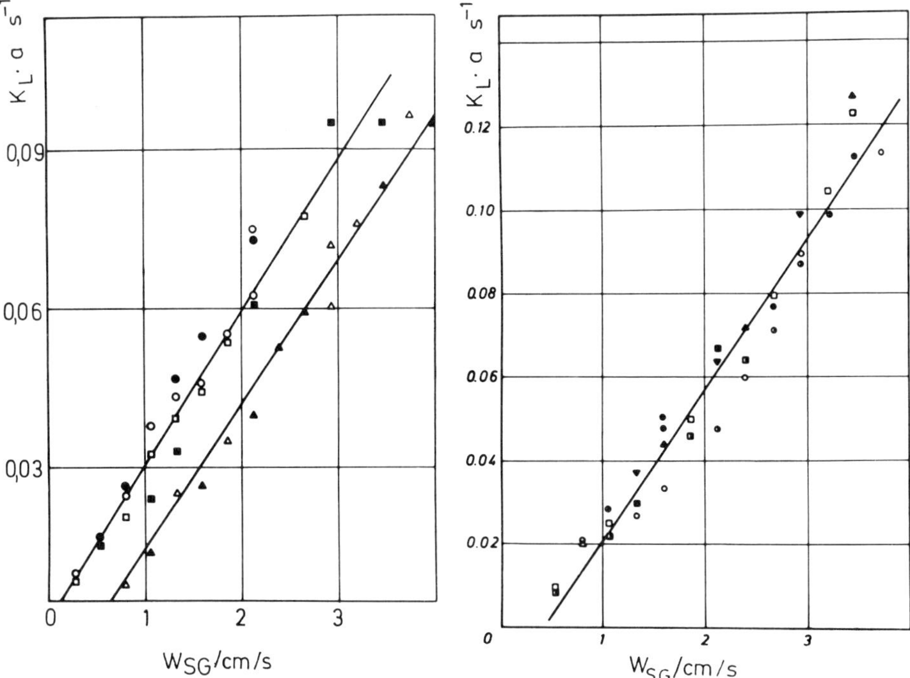

Fig. 62. Typical volumetric mass transfer coefficients (k_La) of oxygen as function of the superficial gas velocity w_{SG} in demineralized water for bubble columns with perforated plate and porous plate gas distributors and for different superficial liquid velocities w_{SL}.

Porous plate	Perforated plate
○ w_{SL} = 0.74 cm/s	△ w_{SL} = 1.69 cm/s
● 1.21	▲ 2.21
□ 1.69	
■ 2.21	

Fig. 63. Typical volumetric mass transfer coefficients (k_La) of oxygen as function of the superficial gas velocity w_{SG} in bubble columns with additives and with perforated plate distributor.

□ 0.5% CH_3OH solution	○ 0.5% C_2H_5OH solution
◨ 1% CH_3OH solution	◑ 1% C_2H_5OH solution
■ 2% CH_3OH solution	● 2% C_2H_5OH solution
▼ 1% CH_3OH solution/NOP	▲ 1% C_2H_5OH/NOP solution

In Fig. 66 data from the present work are compared with the (k_La) values, measured in similar bubble columns by Chang [78] (porous plate, mean pore diameter 175 μm, tap water) and Deckwer [63] (porous plate, mean pore diameter 150 μm, salt solutions and molasses).

The agreement between the (k_La) values for oxygen in water from the present work and those of Chang [78] is fairly good. The difference between the volumetric mass transfer coefficients for oxygen in salt solutions from the present work and those of Deckwer [63] can partly be explained by the different mean pore diameters of porous plates and temperatures of the bubble columns (T = 16 °C (Deckwer), T = 25 °C (present work)). A direct comparison between the glucose system used in the present work

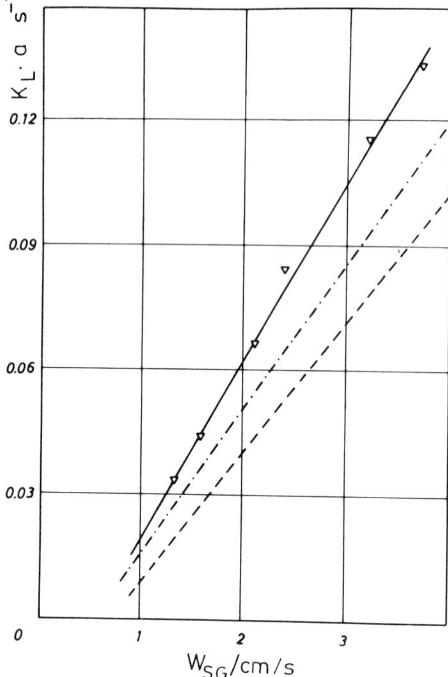

Fig. 64. Comparison of the volumetric mass transfer coefficients ($k_L a$) of oxygen as function of w_{SG} for glucose/perforated plate distributor system with other systems.

▼	2% glucose solutions
–·–·–	alcohols + salts
– – –	H_2O

Fig. 65. Comparison of the volumetric mass transfer coefficients ($k_L a$) of oxygen as function of w_{SG} for different substrates and substrate concentrations.

Porous plate

□	0.5% CH_3OH
◨	1% CH_3OH
■	2% CH_3OH
○	0.5% C_2H_5OH
◑	1% C_2H_5OH
●	2% C_2H_5OH
– – –	perforated plate/alcohols + salts

and the molasses-system of Deckwer is not possible, because the composition of the molasses is not well defined.

The present measurements on the volumetric mass transfer coefficients which occur in bubble columns allow the following conclusions to be drawn: The ($k_L a$) values increase in the investigated range of gas flow rate (0.3 cm/s to 4 cm/s) proportional to the superficial gas velocity for both distributors. The superficial liquid velocity has no effect on ($k_L a$) in the investigated range of 0.74 cm/s to 2.2 cm/s.

With the porous plate distributor higher ($k_L a$) values were achieved than with the perforated plate for pure water as well as for culture media. For the latter the salt and alcohol additives caused a non-specific increase of ($k_L a$). For porous plate systems this

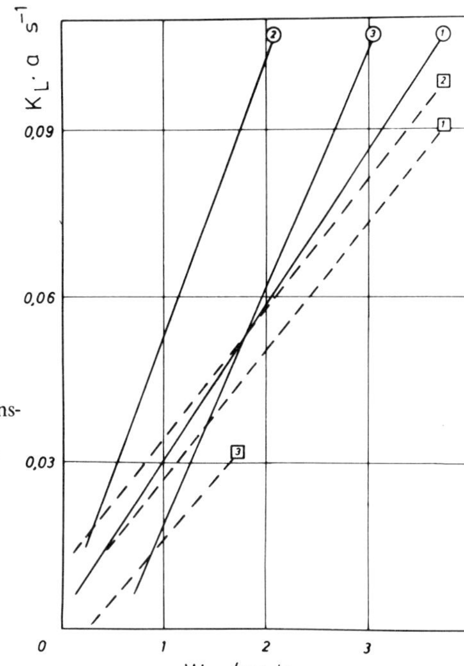

Fig. 66. Comparison of the volumetric mass transfer coefficients (k_La) of oxygen as function of w_{SG} from the present paper with the results of Chang [78] and Deckwer [63].

Porous plate
(1) H_2O — [1] Chang tap water
(2) 0.5% salt — [2] Deckwer (16 °C)
Perforated plate — 0.7 N Na_2SO_4
(3) 2% glucose — 0.17 N NaCl
(1) to (3) — [3] Deckwer
present work — 3.37% molasses

enlargement of (k_La) is specific. At constant gas flow rate the following sequence of increasing (k_La) prevails:

$(H_2O)_1 < (H_2O)_2 \cong (\text{alcohols/salts})_1 < (2\%\ \text{glucose})_1 < (\text{salts})_2 \ll (0.5\%\ CH_3OH)_2 \ll (0.5-2\%\ C_2H_5OH)_2 \cong (1\%\ CH_3OH/\text{salts})_2 < (1\%\ C_2H_5OH/\text{salts})_2$.

Here index 1 refers to the perforated plate and index 2 the porous plate. Fig. 67 shows

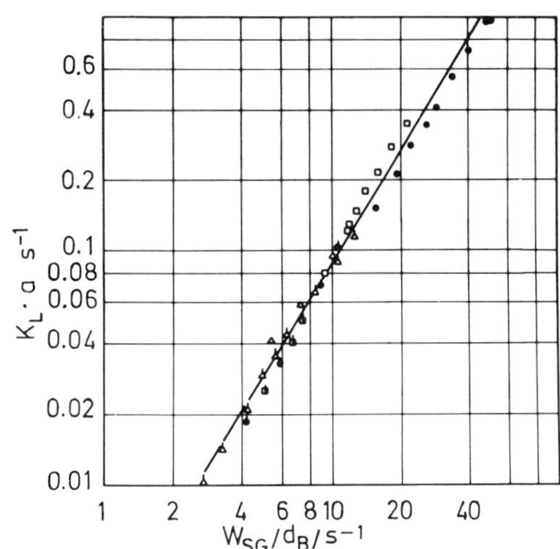

Fig. 67. Correlation of the measured volumetric mass transfer coefficients (k_La) as function of the ratio of superficial gas velocity to the Sauter bubble diameter [based on Eq. (38)].

Perforated plate
△ H_2O
◨ 1% CH_3OH
◑ 1% C_2H_5OH
—— Eq. (38)

Porous plate
△ H_2O
□ 0.5% CH_3OH
◑ 1% C_2H_5OH

that the present data fit a simple regression equation:

$$k_L \cdot a = 0.0023 \left(\frac{w_{SG}}{d_B}\right)^{1.58} \tag{38}$$

with a mean relative error of 10.2%. No useful correlation could be achieved with w_R. By means of the measured $k_L a$ and a-values the mass transfer coefficients k_L can be calculated. Some k_L values are plotted in Fig. 68 as function of w_{SG} for the perforated plate (1) and the porous plate with water (2), 0.5% CH_3OH and/or 1% C_2H_5OH (3) and salts (4). The k_L-values are the largest for the latter. However, the variation of k_L is significantly smaller than the corresponding change of a.

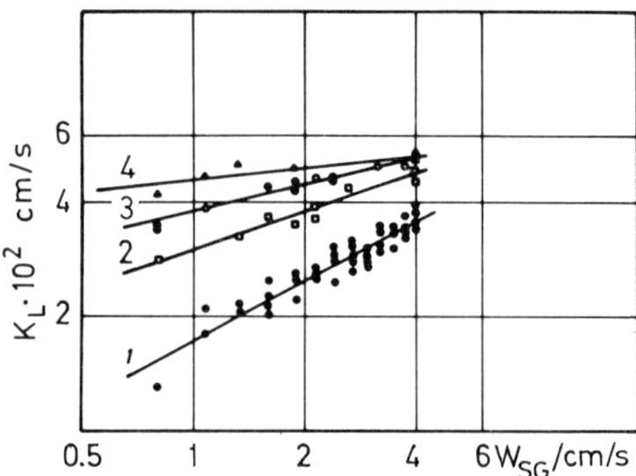

Fig. 68. Comparison of mass transfer coefficients k_L as function of the superficial gas velocity w_{SG} for both the perforated and porous plate. One-stage systems.
Perforated plate
(1) H_2O alcohols and/or salts
Porous plate
(2) H_2O
(3) 0.5% CH_3OH and/or 1% C_2H_5OH
(4) salts

One can recognize that salt and alcohol additives enlarge k_L in bubble columns with the porous plate and have no effect with the perforated plate.

The agreement between the k_L-values (k_L = 0.02 to 0.04 cm/s) of the present work and those of Akita [66] (k_L = 0.02 to 0.03 cm/s) for oxygen and a perforated plate is excellent.

Furthermore, fairly good agreement is found between the k_L-values (k_L = 0.028 to 0.05 cm/s) for the porous plate of the present work and the corresponding data (k_L = 0.04 cm/s) of Forth [79] and Chang [78] for tap water at T = 20–30 °C. In an attempt to obtain more accurate ($k_L a$) values for systems using nozzles the mass transfer coefficient k_L was determined by separate measurements, i.e. in a stirred cell according to Levenspiel [80], and then multiplied by a, which was also obtained by separate measurements. In Fig. 69 k_L is plotted as a function of the stirrer speed (N rpm) for the solutions investigated. One can recognize that k_L is only slightly influenced by the composition of the solution. The measured k_L-values lie in the range 2 to $4 \cdot 10^{-3}$ cm/s for the Reynolds numbers 30–80 ($Re_N = ND_s^2/V_L$, where D_s is the diameter of the stirrer).

A comparison of these values with the ones which were calculated from the $k_L a$ and a obtained in bubble columns, is given in Table 2.

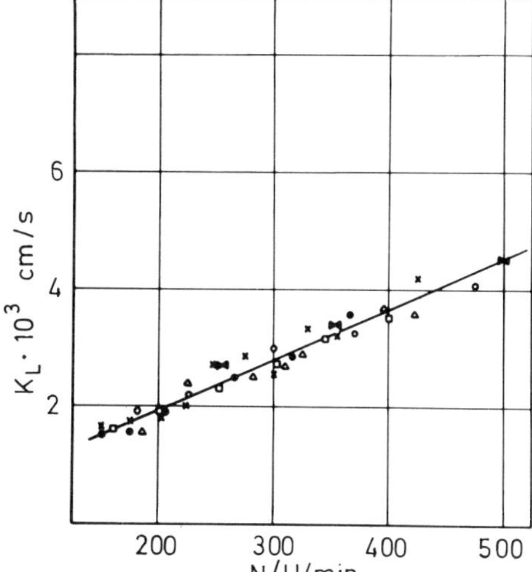

Fig. 69. Coefficient of oxygen mass transfer k_L measured in Levenspiel cell as function of stirrer revolution per minute N in liquid with different composition.
⧓ water (Levenspiel Godfrey [80])
X demineralized water
△ tap water
○ 1% ethanol solution
□ 1% methanol solution
● 1% salt solution.

Table 2. Comparison of k_L values for 1% CH_3OH solution

$k_L \cdot 10^2$ cm/s				
Bubble columns				Levenspiel cell Re_N = 30–80
w_{SG} (cm/s)	perforated plate	porous plate	ejector nozzle	
1.5	2.1	4.2	10	
2.0	2.3	4.5	11.2	0.2–0.4
2.5	3.0	5.0	12.0	

One can recognize that the k_L values which were evaluated using the ejector nozzle, are the highest and, those obtained using the Levenspiel cell, are the lowest. The measured k_L-values for the porous and perforated plates lie inbetween. The differences between the k_L-values evaluated for the same liquids with the ejector nozzle, the porous plate and the Levenspiel cell are considerable. These large differences could result from a number of different factors:

a) Near to the nozzle outlet extremely high specific interfacial areas are produced [81], which diminish with increasing distance from the nozzle outlet due to coalescence processes. The photographic and/or electro optic-methods used lead to the bubble diameters which prevail at some distance from the nozzle and are characteristic of the bubble column. This leads to an overestimate in the value of k_L calculated from $k_L a$ and a.

b) Near to the nozzle outlet high turbulence prevails and bubble coalescence and redistribution occurs, factors which increase the mass transfer coefficient due to interfacial renewal. Even without interfacial renewal the presence of turbulence will increase the mass transfer coefficient [82].

It is not known how large these effects on the mass transfer coefficients are in systems using an ejector nozzle. They are much smaller for the porous plate since the local rate of energy dissipation and by that the local dynamic equilibrium bubble diameter changes only moderately in the column. For the perforated plate both k_L and a are nearly constant, since the bubble diameter is controlled only by the dynamic equilibrium bubble diameter which is nearly constant in the column.

A perforated plate was used in the cell according to the recommendations of Levenspiel [80] to diminish the liquid surface area. Thus small stagnant regions were formed at the liquid surface and the turbulent eddies died away before reaching the interface [83]. Therefore the interfacial renewal is reduced. The low values of k_L in the Levenspiel cell are explained by these phenomena. Therefore the k_L values, evaluated in the Levenspiel cell, cannot be applied to bubble columns and it follows that to estimate k_L requires in situ measurements of both the volumetric mass transfer coefficients k_La and the specific interfacial area a. Furthermore, in order to estimate k_La in bubble columns with very high specific interfacial area new methods must be developed.

The different factors which influence k_L have been discussed by numerous authors (e.g. [31, 53, 66, 84–86]). Most authors have used dimensional analysis to develop equations for k_L. The most important equations having the general form:

$$Sh = f(Re^S, Sc^u) \tag{39}$$

where $Sh = \frac{k_L \cdot d_B}{D_L}$ Sherwood number

$Sc = \frac{\nu_L}{D_L}$ Schmidt-number

$Re = \frac{w_R d_B \rho_L}{\nu_L}$ Reynolds number.

The following regression equation is based on Eq. (39):

$$Sh = 0.15\, Re^{0.75} Sc^{0.5} \tag{40}$$

and is able to describe all fo the present results with a mean relative error of 25.4% (Fig. 70). This standard deviation is small in comparison with the standard deviations of analogous equations (50–100%) given in the literature.

Figure 70 indicates that the Re-numbers in the bubble columns with the porous plate are by a factor of 20 smaller than the corresponding Re-numbers with the perforated plate. In spite of the low Re-numbers (small d_B and w_L) associated with the porous plate system the k_L values are larger than those in the perforated plate system with high Re-numbers (large bubble sizes). This apparent contradiction can be formally explained by the different powers on Re and Sh. The possible physical explanation was given on pages 59/60.

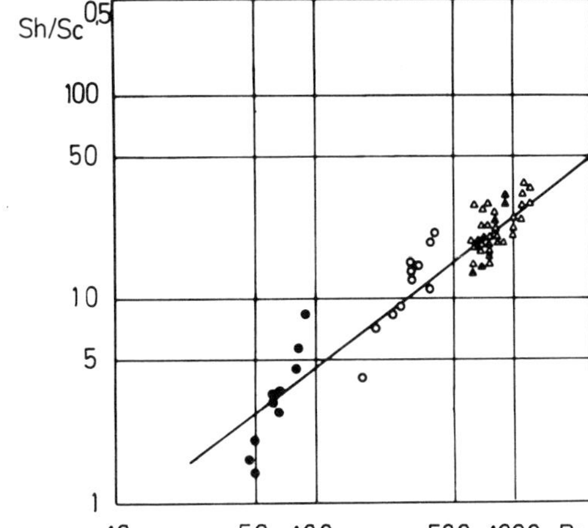

Fig. 70. Correlation between the ratio of Sherwood number to the square root of Schmidt-number and the Reynolds-number.

Perforated plate	Porous plate
△ H_2O, alcohols + salts	▲ H_2O
	○ 0.5% CH_3OH
	● 1.0% C_2H_5OH

Equation (40) is compared on Fig. 71 with the equations recommended by other authors [85, 31, 86, 66]. In the range of high *Re*-numbers (*Re* = 700 to 1150) there is a good agreement between the equation of Reuss [86]:

$$Sh \cdot Sc^{-0.5} = 0.63 \left(\frac{1 - E_G}{1 - E_G^{1/3}}\right)^{0.5} Re^{0.5} \tag{41}$$

and the present data on the perforated plate and the porous plate–water systems. In

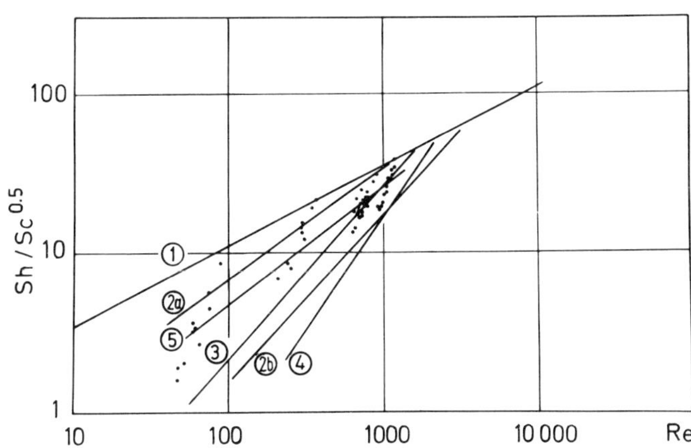

Fig. 71. Comparison of the correlation (4) of the authors with the correlations of other authors.

(1) Higbie [85] [Eq. (45)]
(2a) Calderbank [31] [Eq. (42)]
(2b) Calderbank [31] [Eq. (43)]
(3) Reuss [86] [Eq. (41)]
(4) Akita-Yoshida [66] [Eq. (44)]
(5) present work [Eq. (40)]

the range of intermediate *Re*-numbers ($Re = 200-300$) the equation of Calderbank [31]

$$Sh \cdot Sc^{-0.5} = 0.42 \left(\frac{d_B^2 \Delta\rho g}{\nu_L \rho_L w_R}\right)^{0.33} Re^{0.33} \tag{42}$$

agrees fairly well with the results obtained for 0.5% CH_3OH with the porous plate. Equation (42) is valid for $d_B > 2.5$ mm, but for $d_B < 2.5$ mm Calderbank recommends Eq. (43):

$$Sh \cdot Sc^{-0.5} = 0.31\, Sc^{-7/6} \left(\frac{\Delta\rho \nu_g w_R^3}{\rho}\right)^{0.33} Re. \tag{43}$$

The Eq. (43) as well as Eq. (44) of Akita [66]:

$$Sh \cdot Sc^{-0.5} = 0.5 \left(\frac{g\, d_B^2 \rho_L}{\sigma}\right)^{3/8} \left(\frac{g\, d_B^2}{V_L w_R}\right) Re^{0.25} \tag{44}$$

yield low *Sh*-numbers in the range $Re = 50-1150$. In contrast Eq. (45):

$$Sh \cdot Sc^{-0.5} = 1.13\, Re^{0.5}, \tag{45}$$

which was recommended by Higbie [85], covers the entire range of *Re*-numbers but predicts too high *Sh*-numbers probably because of the use of the simple contact time $\theta = \frac{d_B}{w_R}$, which only considers the one-dimensional movement of the bubbles and leads to an underestimate of contact time.

All of the above mentioned equations contain a simple *Re*-number, which alone cannot characterize sufficiently the fluid dynamic state of the bubble swarm. This is also true for the recommended Eq. (40), which describes the performance of the systems investigated fairly well, but is also of limited applicability.

h) Back Mixing in the Liquid Phase

Since the bubble column used was relatively short [100] and the two phase flow changed its character from laminar to turbulent in about half the height of the column, the measurements were carried out only in the lower half of the column, i.e. from the distributor up to the position of tracer injection (175 cm from the distributor). Since the axial position of the flow transition depends on the operating parameters and the exit section has a very complicated flow structure, no measurements were made in the upper half of the column. Therefore, the results presented here, are characteristic of the gas entrance section and cannot be applied to tall bubble columns.

Three different distributors were used: the perforated plate already described and two porous plates with mean pore diameters of 50 and 5 μm.

Figure 72 shows the coefficient of back mixing D_{LB} in the liquid phase as function of the superficial gas velocity for the perforated plate distributor.

A comparison of Fig. 72 with Fig. 30 indicates that in alcohol systems coalescence is supressed. In such systems the mixing coefficient D_{LB} remains very low in the investi-

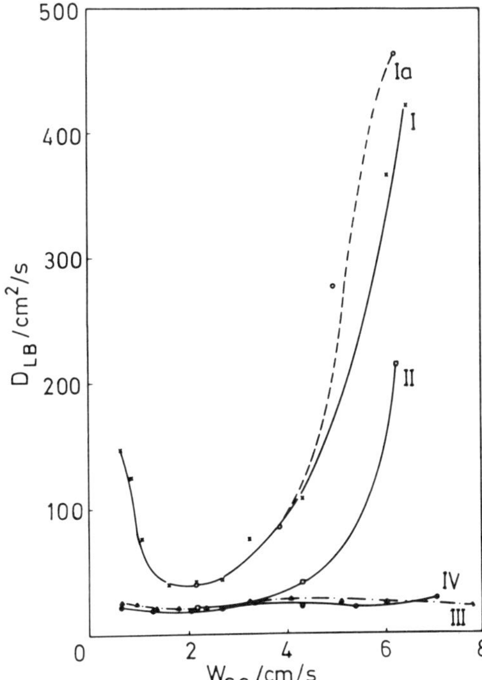

Fig. 72. Coefficient of backmixing in the liquid phase, D_{LB}, as function of the superficial gas velocity. Perforated plate aerator. ○ I a, × I tap water, □ II 1% salt solution △ III 2% methanol solution, ● IV 0.5% ethanol solution. (I a dye tracer)

gated range of w_{SG}. With salt solution the coalescence is retarded but not completely supressed. Therefore a steep increase of D_{LB} appears at high superficial gas velocities at which the relative swarm velocity w_R also increases. In tap water the coalescence rate is high. Therefore w_R and D_{LB} begin to increase at relatively low superficial gas velocities. When using tap water with porous plates there is a rapid increase in D_{LB} as w_{SG} is increased (Fig. 73, 74). With methanol and up to w_{SG} = 4 cm/s, the mixing coefficient is again extremely low. However, at high superficial gas velocities a rapid increase in D_{LB} occurs (Fig. 73). The transition into the turbulent state is shifted to higher gas velocities in comparison with the water system. Similar behaviour was found with the 5 μm pore diameter porous plate (Fig. 74). However, for methanol the transition region was shifted to gas velocities outside the operating range of the apparatus.
The behaviour of D_{LB} for n-propanol and porous plates (Figs. 73, 74) is unexpected: fairly high mixing coefficients were found at low superficial gas velocities, particularly with the porous plate with the pore diameter of 5 μm. The initial bubble diameters are very small with these distributors and because of the very effective suppression of the coalescence by n-propanol the original bubble size is nearly preserved. Thus the buoyancy forces are very small and the relative swarm velocity is very low and changes only slightly with the superficial gas velocity, i.e. it is similar to the behaviour of ejector and injector nozzles [curves (4) and (5) in Figs. 33 and 34]. Therefore with increasing gas throughput the mean relative gas hold-up increases up to 0.5 (for the plate with a

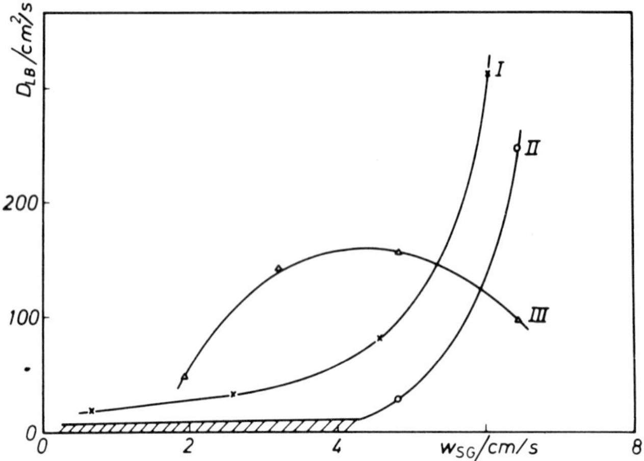

Fig. 73. Coefficient of backmixing in the liquid phase, D_{LB}, as function of the superficial gas velocity. Porous plate aerator with mean pore diameter fo 50 microns. × tap water, ○ 0.5% methanol solution, △ 0.5% n-propanol solution.

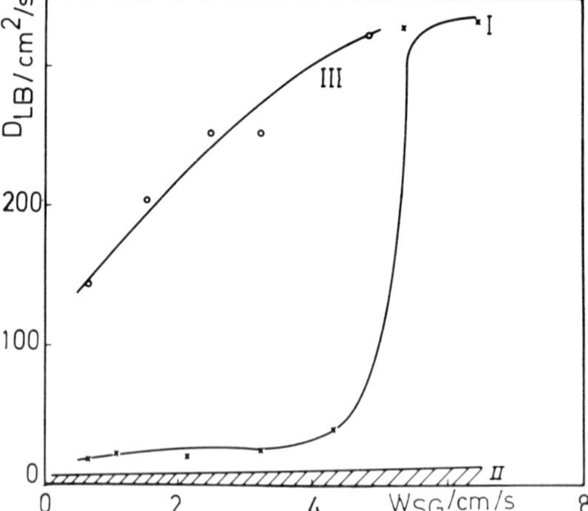

Fig. 74. Coefficient of backmixing in the liquid phase, D_{LB}, as function of the superficial gas velocity. Porous plate aerator with mean pore diameter of 5 microns. × I tap water, II 0.5% methanol solution. ○ III 0.5% propanol solution.

50 μm pore diameter) and to 0.62 (for the plate with a 5 μm pore diameter) and the cross-section of the liquid flow significantly diminishes. Hence, the effective linear liquid velocity becomes very high.

In general, at high enough superficial gas velocities turbulence is induced by the bubble formation at the gas distributor. However, this turbulence decays quickly because it cannot be supported by the flow (the liquid *Re*-number is too low). With increasing distance from the gas distributor bubble coalescence causes a non-uniform increase in the bubble size. Because of the diminishing liquid flow cross section (increasing linear velocity) and the non-uniform bubble velocities the two phase flow becomes unstable and transition into the turbulent state occurs. In bubble columns with the porous plate of 5 μm pore diameter and n-propanol solution the turbulence induced at the gas distributor does not decay but is preserved due to the high liquid velocities (high liquid *Re*-numbers).

Because of the complexity of this phenomenon no quantitative relation exists to calculate D_{LB}. The relation (46) cannot be applied, because it was measured in the middle section of the bubble column and therefore it does not include the entrance effect. The relation of Deckwer [63] cannot be applied either because his columns were taller, hence the intensity of mixing was much higher due to the dominating influence of the turbulence above the entrance region.

6. Multistage Systems

a) Mean Relative Gas Hold-Up and Relative Gas Velocity

For multistage systems a great number of combinations of the most important construction parameters is possible. In the present work only few were varied, namely,
number of stages: 6 or 12
length of compartments: 39 cm (for 6 stages) or
19.5 cm (for 12 stages),
separation of the compartments by perforated plates:
plate A with free surface area of 28% and hole diameter of 0.2 cm, and
plate B with free surface area of 12.5% and hole diameter of 0.4 cm.
The influence of the number and length of stages and the properties of the perforated plates on the behaviour of the column is complex. Since the bubble size plays a decisive role with regard to both the relative mean hold-up and the specific interfacial area, the influence of the compartments on the bubble size has to be considered. By the use of perforated plates to separate compartments the bubble size distribution can be influenced in various ways:

a) If the redistribution of the gas phase is similar to the gas distribution in the "bubbling gas" range, large free surface area and small orifice diameters are preferable.
b) If a "gas jet" is formed in the laminar liquid, small orifice diameters yield small bubbles, but the free surface area does not influence the bubble size.
c) If the gas jet is broken-up by the local turbulent field at the plate, a small free surface area (high intensity of turbulence) is preferred. By diminishing the orifice diameters the scale of the turbulence is reduced. This yields high local energy dissipation densities at the plates, and this decays quickly with increasing distance.

In general the intensity of the small scale turbulence is increased in a multistage system at the expense of the large scale turbulence. Since the small scale turbulence dissipates the energy with much higher efficiency than the large scale turbulence, one produces by the former a higher rate of local energy dissipation than by the latter. However, this high energy dissipation rate decays much quicker than the one due to larger scale turbulence [91]. In contrast to single stage systems with a porous plate distributor in which the rate of the energy dissipation only slightly changes along the column, in multistage systems using perforated plates, the rate of the energy dissipation locally increases at each stage separator but quickly diminishes with increasing distance from the separator plate.

By appropriate variation of the hole diameter of the perforated plate and/or its free surface area, the scale and/or the intensity of the turbulence and by that the rate of local energy dissipation can be adjusted in the column according to the properties of the fermentation medium, e.g.

α) By applying a coalescing system (e.g. pure water) the local bubble size is determined by local dynamic equilibrium bubble diameter. Since the energy dissipation is mainly caused by small scale turbulence the dynamical equilibrium bubble diameter will periodically decrease at every perforated plate but quickly increase again. The smaller the distance between the plates the higher the average level of the energy dissipation density and the smaller the average dynamical equilibrium bubble diameter which can be maintained. Therefore one would expect for water, that the relative mean gas holdup would increase as follows: one-stage < six-stage < twelve-stage.

β) For non coalescing systems the dynamical equilibrium diameter in the column (far enough from the perforated plate) plays no important role, i.e. only the bubble diameter at the plate is of importance.
Under the conditions applied in these investigations it is expected that the redistribution of the gas phase occurs according to mechanism c above. Hence, the higher the rate of local energy dissipation at the plate (caused by small orifice diameter and small free surface area), the smaller is the bubble size. This bubble diameter is largely preserved in the rest of the compartments.

γ) For intermediate and weakly coalescing systems there can be a wide spectrum of behaviour depending on the relative importance of the influence of the various parameters.

The experimental results on the effect of the solute on E_G are similar to those obtained for the single stage systems. E_G becomes larger with alcohol addition. This effect being more significant for high alcohol homologues (Fig. 75). In the coalescing system (pure water) the multistage columns always produced a higher E_G than the corresponding single stage column (Figs. 76–80).

In the six-stage column the perforated plates A and B produce nearly the same E_G in water (Fig. 80), i.e. the favourable higher wave number (smaller scale) of turbulence due to the smaller hole diameter is compensated by the unfavorable lower intensity of turbulence due to the higher free surface area (plate A) and/or vice versa (plate B).

In twelve-stage solumn plate B produced higher E_G than plate A, i.e. the higher intensity of turbulence overcame the influence of the lower wave number (Fig. 79).

With plate B there is no difference between the E_G of the 6- and 12 stage columns with water (Fig. 77) in contrast with plate A (Fig. 78), which produces higher E_G for the 6-stage than for the 12-stage column if $w_{SG} > 2$ cm/s. This is unexpected if one does not consider the possible coalescence promoting effect of the separating plates at high gas flow rates. This effect is larger for plate A than for B (see also the behaviour of w_R).

For a nearly non-coalescing system (ethanol), if the turbulence did not play any role, plate A would produce higher E_G than plate B. However, this is not the case, neither for the 6-stage (Fig. 80) nor for the 12-stage units (Fig. 79). Hence, one can conclude that the initial bubble diameter is controlled by the local turbulence at the separating plates. Ethanol solution in a multistage system always produced higher mean relative gas hold-up than in the single stage system (e.g. Fig. 76). For ethanol solutions with

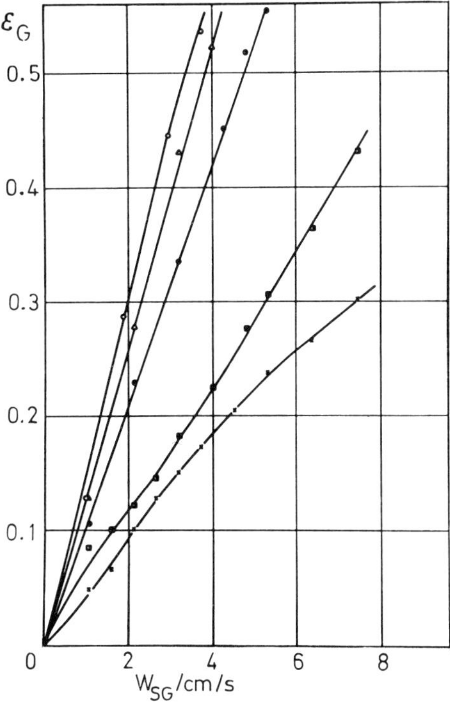

Fig. 75. Influence of the alcohol solutes on E_G. Six-stage cascade. Plate A (with 28.5% free surface area and hole diameter of 0.2 cm). Porous plate gas distributor.
× H_2O
◧ 1% CH_3OH
◑ 1% C_2H_5OH
△ 1% n-C_3H_7OH
○ 1% n-C_4H_9OH

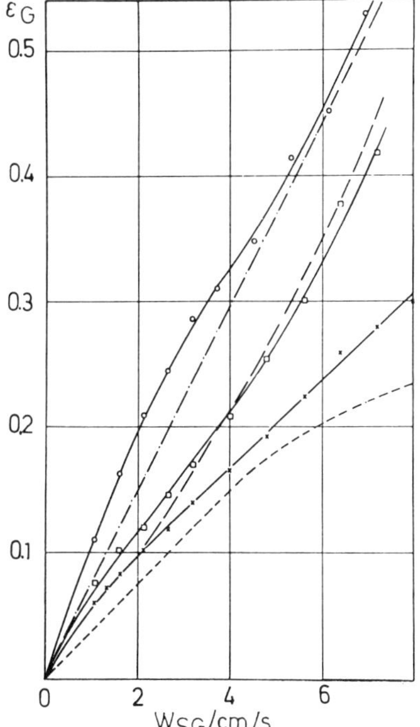

Fig. 76. Comparison of E_G of single and twelve-stage bubble columns with different solutes. Perforated plates with 28.5% free surface area and with 0.2 cm hole diameter. Porous plate gas distributor. H = 334 cm, T = 25 °C
× H_2O, ○ 1% C_2H_5OH, □ 1% CH_3OH } 12-stage
—— CH_3OH single-stage
–·–· C_2H_5 single-stage
– – – H_2O, single-stage

plate B no difference occurred between the E_G's of 6- and 12-stage columns (Fig. 77), in contrast to plate A, which produces higher E_G for the 6-stage than for the 12-stage columns, if $w_{SG} > 2$ cm/s (Fig. 78). This behaviour is analogous to that of the coalescing (water) system and can be explained again by the stronger coalescence promoting effect of plate A compared with plate B at higher flow rates.

1% methanol solution is considered an intermediate coalescing system. It is interesting to note that in methanol systems no differences were found between the mean relative gas hold-ups for 1, 6 and 12-stages (Figs. 76–78) or with plates A and B (Figs. 79 and 80). This phenomenon is not understood yet. The terminal velocities w_T, i.e. the relative velocities w_R for $w_{SG} = 0$, indicate that the residence time of the gas phase is significantly higher (w_T is smaller) in a multistage system compared with a single stage system. Thus for 1% C_2H_5OH the terminal velocity is diminished from w_T = 21 cm/s (single-stage) to

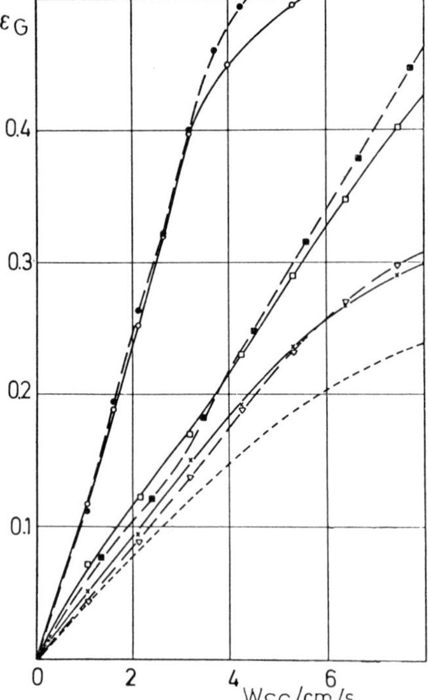

Fig. 77. Comparison of gas hold ups ε_G for six- and twelve-stage columns. Plate B (free surface area: 12.5%; hole diameter: 0.4 cm). Gas distributor: porous plate. H = 334 cm, T = 25 °C

6-stage	12-stage	
×	▾	H_2O
□	■	1% CH_3OH
○	●	1% C_2H_5OH
– – –		H_2O single-stage

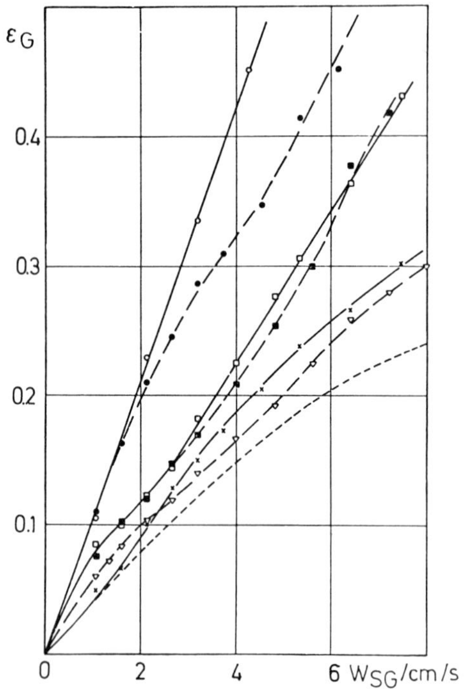

Fig. 78. Comparison of gas hold ups, ε_G, for six- and twelve-stage columns. Plate A (free surface area: 28.5%; hole diameter: 0.2 cm). Gas distributor: porous plate, H = 334 cm, T = 25 °C. (for symbols see Fig. 77)

w_T = 7 cm/s [six-stage (Fig. 81) and twelve-stage (Fig. 82)]. This increase in the residence time of the gas phase is due to the retardation of the bubbles by the separating plates. The retardation of the bubbles is partly compensated, especially at higher superficial gas velocities, by the increased coalescence rate due to the plates. This effect seems to be smaller for plate B than for plate A [e.g ethanol, water in the twelve-stage system

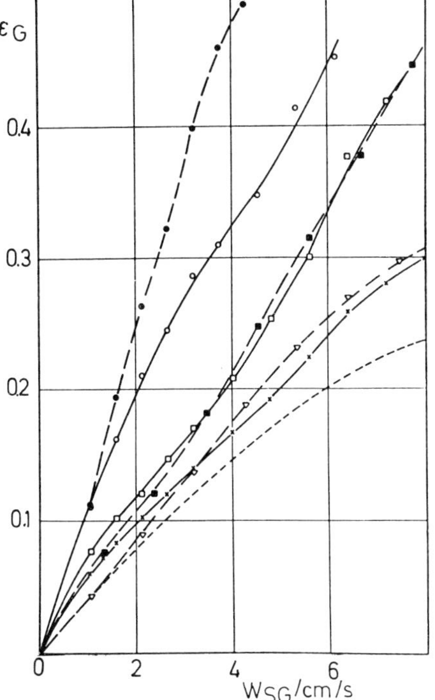

Fig. 79. Comparison of gas hold ups, E_G, for twelve-stage columns with different free surface areas and hole diameters of the perforated plate. Gas distributor: porous plate. H = 334 cm, T = 25 °C.

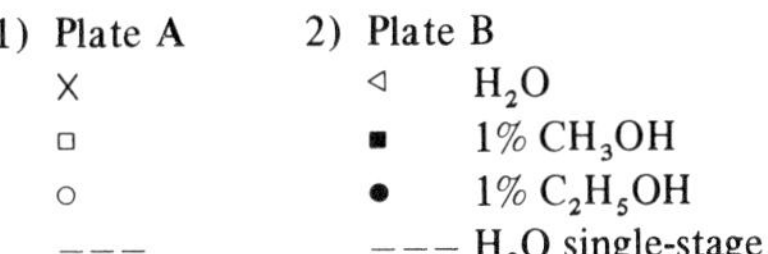

1) Plate A	2) Plate B	
×	◁	H_2O
□	■	1% CH_3OH
○	●	1% C_2H_5OH
– – –	– – –	H_2O single-stage

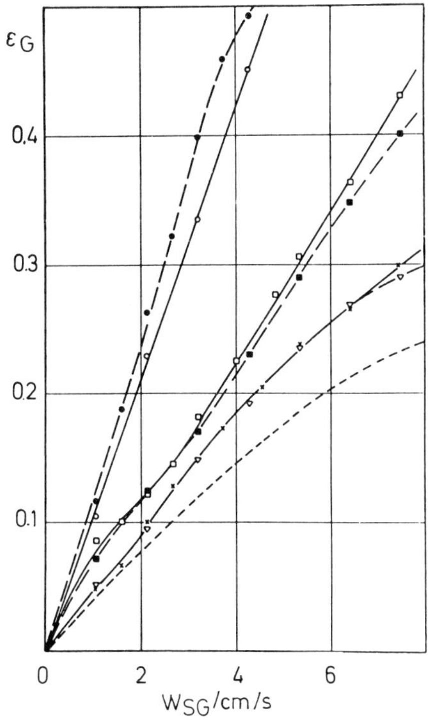

Fig. 80. Comparison of gas hold ups, E_G, for six-stage columns with different free surface areas and hole diameters of the perforated plates. Gas distributor: porous plate. H = 334 cm, T = 25 °C.

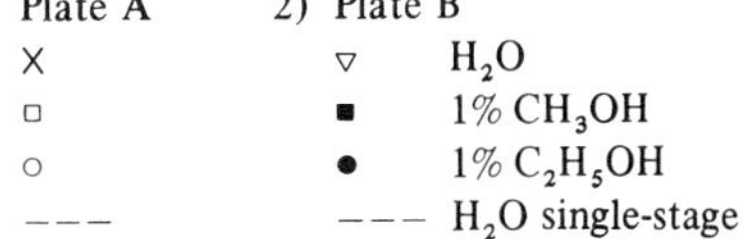

1) Plate A	2) Plate B	
×	▽	H_2O
□	■	1% CH_3OH
○	●	1% C_2H_5OH
– – –	– – –	H_2O single-stage

(Fig. 82)] if the superficial gas velocity w_{SG} is intermediate or high. The curves $w_R(w_{SG})$, are not completely understood and additional data are needed to understand fully the behaviour of multistage systems. However, these results show that multistage systems can be designed to accommodate the fluid properties and lead to advantageous characteristics.

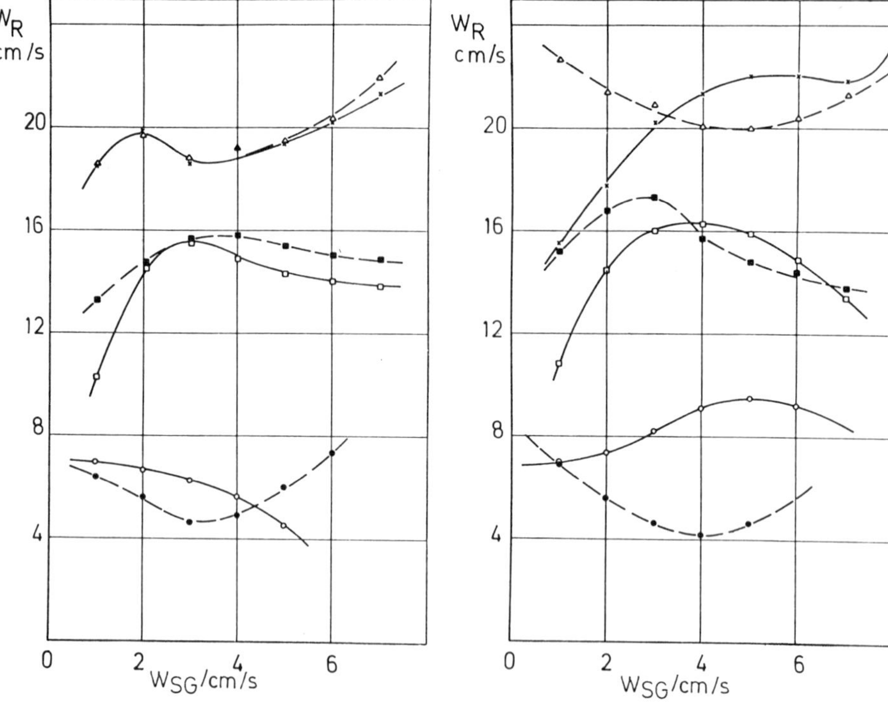

Fig. 81. Relative gas velocity w_R as function of the superficial gas velocity w_{SG} in six-stage bubble column. Porous plate gas distributor. w_{SL} = 2.2 cm/s.
5 perforated plates with

Plate A	Plate B	
×	△	H_2O
□	■	1% CH_3OH
○	●	1% C_2H_5OH

Fig. 82. Relative gas velocity w_R as function of the superficial gas velocity w_{SG} in twelve-stage bubble columns. Porous plate gas distributor. w_{SL} = 2.2 cm/s.
11 perforated plates with

Plate A	Plate B	
×	△	H_2O
□	■	1% CH_3OH
○	●	1% C_2H_5OH

b) Volumetric Mass Transfer Coefficient

Because of the coalescence and redispersion of the bubbles during their passage through the perforated plates, surface renewal is more significant in multistage columns than in a single stage column. The influence of the construction parameters of multistage equipment on surface renewal is not known. For pure water the difference between the volumetric mass transfer coefficients (k_La) of oxygen in single-, six- and twelve-stage systems is relatively slight (Figs. 83 and 85). With plate A (k_La) increases in the following sequence: single-stage < twelve-stage < six-stage (Fig. 83) and with plate B: single-stage

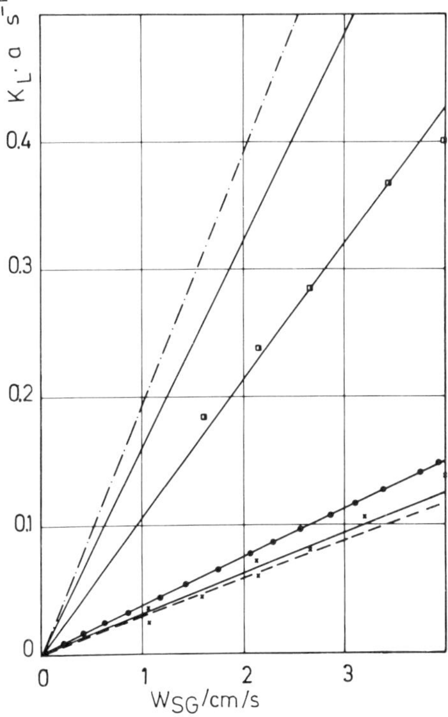

Fig. 83. Comparison of the volumetric mass transfer coefficients ($k_L a$) for oxygen in one-, six- and twelve-stage bubble columns with H_2O and/or 1% CH_3OH solute and plate A (with 28.5% free surface area and 0.2 cm hole diameter). w_{SL} = 2.2 cm/s. Porous plate gas distributor.

-.-.- single-stage-system 1% CH_3OH;
--- H_2O
—— six-stage system } 1% CH_3OH
▫ twelve-stage system }
•-•-• } six-stage system } H_2O
X } twelve-stage system }

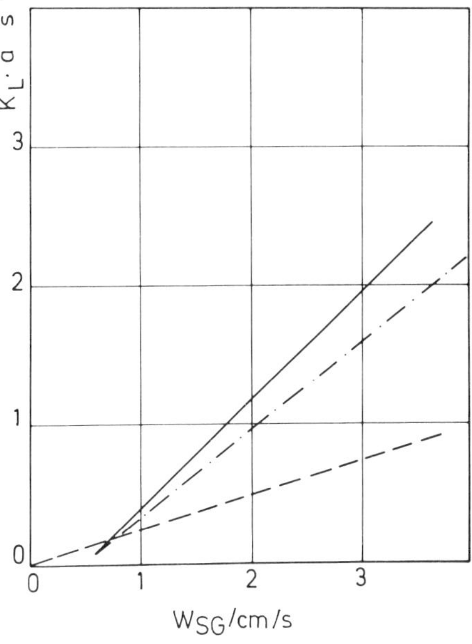

Fig. 84. Comparison of the volumetric mass transfer coefficients ($k_L a$) for oxygen in one-, six- and twelve-stage bubble columns with 1% C_2H_5OH solute and plate A. w_{SL} = 2.2 cm/s. Porous plate gas distributor.

--- single-stage system
—— six-stage system
-.-.- twelve-stage system

< six-stage < twelve-stage (Fig. 85). This can be explained again by the higher coalescence promoting effect of plate A than that of plate B. With methanol solute, the differences in $k_L a$ are more significant than for water, i.e. $k_L a$ decreases in following sequence: single-stage > six-stage > twelve-stage for plate A (Fig. 83) and twelve-stage > single-stage > six-stage for plate B (Fig. 85).

The highest volumetric mass transfer coefficients are up to 3 s^{-1} and were achieved using ethanol in a six-stage system with plate A (Fig. 84).

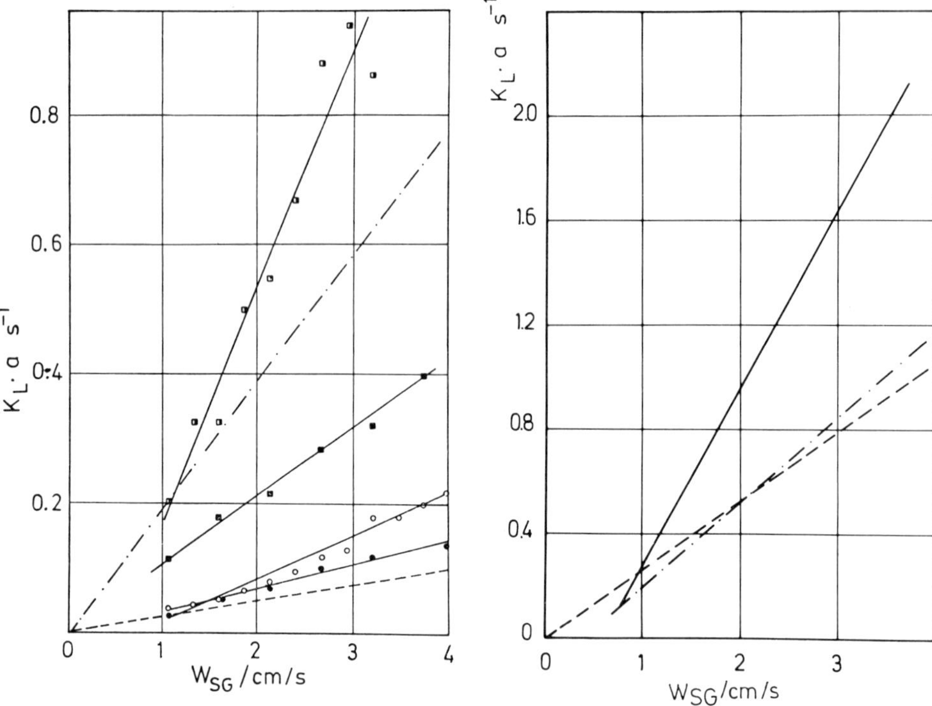

Fig. 85. Comparison of the volumetric mass transfer coefficients ($k_L a$) for oxygen in single-, six- and twelve-stage bubble columns with H_2O and/or 1% CH_3OH solute and plate B (with 12.5% free surface area and 0.4 cm hole diameter). w_{SL} = 2.2 cm/s. Porous plate gas distributor.

---	single-stage system	H_2O
●	six-stage system	
○	twelve-stage system	
–·–·–	single-stage system	1% CH_3OH
■	six-stage system	
◘	twelve-stage system	

Fig. 86. Comparison of the volumetric mass transfer coefficient ($k_L a$) for oxygen in single-, six and twelve-stage bubble columns with 1% C_2H_5OH solute and plate B. w_{SL} = 2.2 cm/s. Porous plate gas distributor.

--- single-stage system
–·–·– six-stage system
—— twelve-stage system

For ethanol and the same plate $k_L a$ decreased in the sequence: six-stage > twelve-stage > single-stage.

The difference between the $k_L a$-values in single and multistage systems with ethanol solute and with plate B is relatively small, ($k_L a$) decreased in the sequence: twelve-stage > six-stage > single-stage (Fig. 86).

Thus with six-stage columns plate A and with twelve-stage columns plate B produces the higher ($k_L a$) values.

It is remarkable that the bubble velocities in the six-stage bubble column differ only moderately between plates A and B (Fig. 81), but their ($k_L a$)-values are very different, i.e. ($k_L a$) = 1.9 s^{-1} with plate A and ($k_L a$) = 0.9 s^{-1} with plate B, under the same conditions of ethanol solution and w_{SG} = 3 cm/s.

Clearly $(k_L a)$ is a complex function of the construction and process parameters in a multistage system.

c) Longitudinal Dispersion in the Liquid Phase

In addition to the single and six-stage columns operated with two countercurrent phases, a countercurrent single-stage three-phase column was used. The latter contained hollow glass beads (Microballons FT 102, Emerson and Cumming Inc., Mass., U.S.A.) 125–250 μm in diameter, with a mean density of 0.28 g/cm^3. The three-phase bubble column was investigated as a possible fermentation system containing inert solids to increase the gas hold-up and/or to serve as a carrier for the microorganisms.
The details of the apertures were as follows:

Table 3

No.	Phases	Operation	Stages	Height/cm	Diameter/cm
1	2	concurrent	1	440	14
2	2	concurrent	6	440	14
3	2	countercurrent	1	380	14
4	3	countercurrent	1	380	14

Axial mixing was measured in the liquid phase by the pulse tracer technique using 20% NaCl solution. The tracer concentrations were measured at two locations in the column and stored by data logger for subsequent computer by fitting of the calculated system transfer function Eq. (24) to the measured data by means of the non-linear optimization of Eq. (26). The standard deviations of the fitted Peclet number were of order 10^{-3}.
Figures 87 and 88 show the effect of the superficial gas velocity on the Peclet number in various systems. For all the systems investigated the axial dispersion increases (the Peclet number decreases) with increasing liquid and superficial gas velocities. The intensity of longitudinal mixing is lower in multistage bubble columns than in other systems at the same gas rate.
The following empirical correlation describes the fluid mixing in the systems investigated:

$$Pe^* \cdot w_L = B_0 \left(\frac{w_{SG}}{w_{SL}}\right)^{B_1} \tag{46}$$

with the values of parameters given in Table 4.

Table 4. Empirical parameters B_0, B_1 of Eq. (46)

No.	B_0	B_1
1	14.57 ± 0.42	−0.589 ± 0.081
2	24.94 ± 0.60	−0.562 ± 0.105
3	6.014 ± 0.154	−0.307 ± 0.034
4	2.551 ± 0.061	−0.0044 ± 0.0003

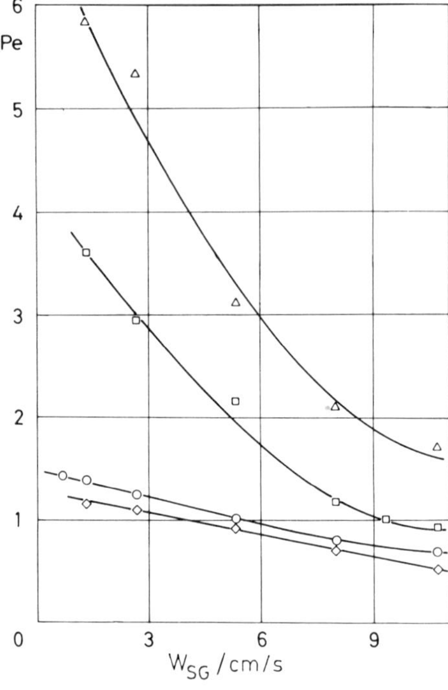

Fig. 87. Effect of the superficial gas velocity on the *Pe*-number, two-phase systems. w_{SL} = 1.2 cm/s
countercurrent: single-stage ○
concurrent: single-stage □
six-stage △
Three-phase system
countercurrent: single-stage ◇

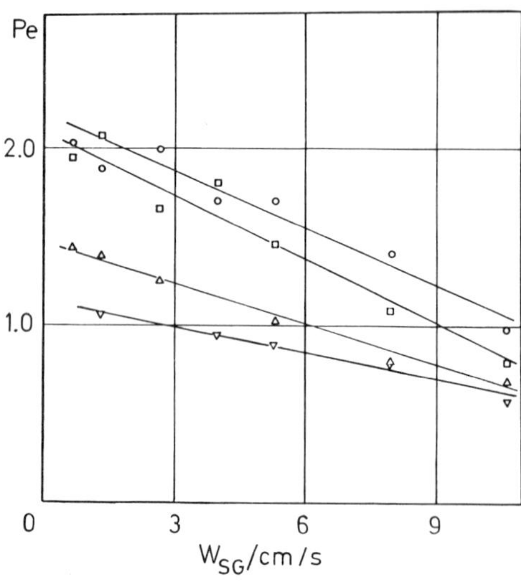

Fig. 88. Effect of the superficial gas velocity on the *Pe*-number for different superficial liquid velocities.
two-phase countercurrent, one stage
w_{SL} = 2.38 cm/s ○ 1.20 cm/s △
1.68 cm/s □ 0.71 cm/s ▽

Figure 89 shows the data $Pe^* w_L$ compared with values calculated from Eq. (46). These results are compared with literature data on axial mixing, by using the w_{SG}/w_{SL}-values of the published data [87–90]. $Pe^* \cdot w_L$ was calculated by Eq. (46) and B_0 and B_1 values of Table 4 (Fig. 90). Most of the data fall within the dashed lines which represent a mean relative error of ±20%.

d) Conclusion

The residence time of the bubbles is much higher in multistage-bubble columns than in single stage columns. This effect which increases the gas gold-up, is partly compensated by the higher coalescence rate of the bubbles, especially at higher gas flow rates. In six-stage columns the application of plates A (larger free surface area and smaller hole diameter) and in twelve-stage columns the application of plates B (smaller free surface area and larger hole diameter) yields the better performance.

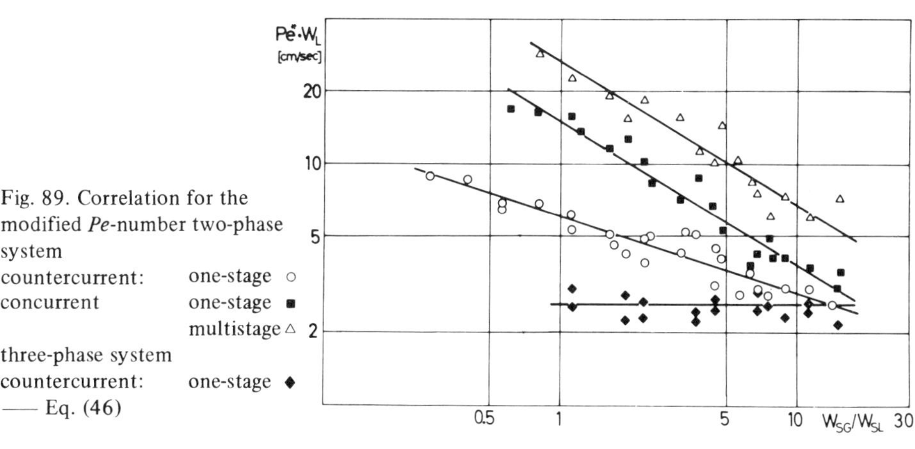

Fig. 89. Correlation for the modified *Pe*-number two-phase system
countercurrent: one-stage ○
concurrent one-stage ■
multistage △
three-phase system
countercurrent: one-stage ◆
—— Eq. (46)

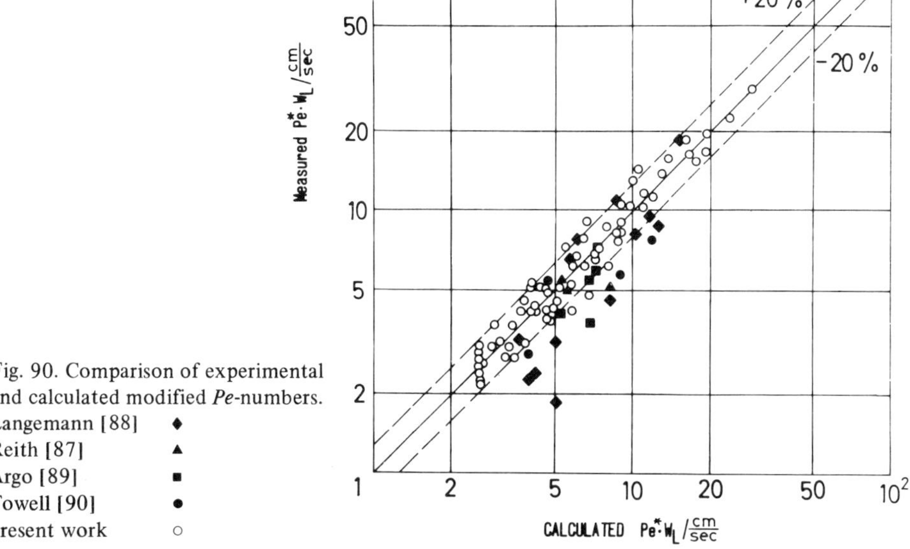

Fig. 90. Comparison of experimental and calculated modified *Pe*-numbers.
Langemann [88] ◆
Reith [87] ▲
Argo [89] ■
Towell [90] ●
present work ○

Multistage systems seem to be very flexible not only with regard to the residence times distribution of the gas and liquid phase but also the relative gas velocity (mean residence time of the bubbles in the column).

By applying suitable multistage systems the volumetric mass transfer coefficient can be significantly increased and the longitudinal diffusion coefficient can be considerably reduced.

7. Comparison of Bubble Columns with Air Lift Fermentors and Mechanically Agitated Fermentors

In bioreactors for aerobic processes, for high growth rate of the microorganisms, and for high cell concentrations, the effective growth rate is determined by the oxygen transfer rate which is controlled by the specific gas/liquid interfacial area. In Table 5 the specific surface areas which are achieved in different bioreactor types under comparable (sulphite oxidation) conditions, are compared as a function of the specific power input (E/V). As can be seen from the table, mechanically agitated bioreactors, MAF, yield the smallest specific areas with the highest power input, while Frings bioreactor with its self-aspirating aerator is the best of this type. The latter is better than the air-lift bioreactors with draught tube, ALF. The highest specific surface areas are obtained by bubble column bioreactors, BCF, with ejector and/or

Table 5. Comparison of specific surface areas in various fermentor types measured under sulphite oxidation conditions (for symbols see below)

Fermentor type	Aerator type	A m^{-1}	E/V kW/m^3	w_{SG} cm/s	Literature
MAF		120[a]	1		[105]
		300	1		[77]
		400	1		[77]
		600	2		[77]
BCF	PE	650	0.6	4.5 (0.63 vvm)	(authors)
FF		1000	1.2		[103]
MAF		1000	4		[77]
		1100[a]	10		[105]
ALF	EN	1300	0.9	3	[104]
		1300	0.9	8	[104]
		1700	1.8	3	[104]
		1800	1.8	8	[104]
		2000	7.2	3	[104]
BCF	PO	2000	0.9	4 (0.56 vvm)	(authors)
MAF		2200	10		[77]
FF		2315	2.77	(0.48 vvm)	[103]
ALF	EN	2500	7.2	8	[104]
BCF	IN	6000	1.5	3 (0.42 vvm)	(authors)
BCF	EN	8000	2.2	7 (1.4 vvm)	(authors)

MAF mechanically agitated reactor (stirred tank)
ALF air lift reactor with draught tube
BCF bubble column reactor
FF "Frings fermentor" (mechanically agitated)
PE perforated plate aerator
PO porous plate aerator
IN injector nozzle aerator
EN ejector nozzle aerator

[a] Measured by the reaction: $NaOH + CO_2$

injector nozzle aerators. A comparison of the volumetric mass transfer coefficients ($k_L a$) which can be achieved in different bioreactor types (Table 6) also support his conclusion. Bubble column reactors with non-coalescing media yield volumetric mass transfer coefficients, ($k_L a$), which are by order of magnitude higher than those in the conventional mechanically agitated fermentors.

One can recognize from this comparison in Tables 5 and 6 that bubble columns with porous plate aerators are very economical reactors, since they yield medium to high specific surface areas with relatively low specific power input. Bubble column reactors with either an injector or ejector nozzle are also highly effective bioreactors for aerobic processes, because they produce extremely high specific surface areas with intermediate power inputs.

Table 6. Comparison of the volumetric mass transfer coefficients measured in various bioreactors (for symbols see Table 5)

Fermentor type	Aerator type	$k_L a$ s^{-1}	E/V kW/m^3	w_{SG} cm/s	Literature
MAF		0.083	1	(0.5 vvm)	[107] $V = 51\,000$ l
		0.083	1	(1.1 vvm)	[107] $V = 3000$ l
		0.083	1	(0.8 vvm)	[107] $V = 500$ l
BCF	PE	0.09	0.5	3 (0.6 vvm)	(authors) alcohol + salt
		0.03[a]	0.71	6.3 (0.88 vvm)	alcohol + salt [100]
MAF		0.11	1	(3.5 vvm)	[107] $V = 200$ l
MAF		0.11–0.22[a]	1–3		[106] up to $V = 100$ m³
BCF	PO	0.9	0.9	4 (0.8 vvm)	(authors) methanol + salt
		0.138[a]	0.43	3.8 (0.53 vvm)	methanol + salt [100]
BCF	PO	1.25	0.8	3 (0.6 vvm)	(authors) ethanol + salt
		0.320[a]	0.26	2.3 (0.32 vvm)	ethanol + salt [100]

[a] During cell growth

8. Economic Outlook

Under oxygen-limiting conditions the relationship between dissolved oxygen and cell concentration can be predicted in well mixed continuous culture systems assuming the applicability of Monod's model:

$$\frac{dx}{dt} = \frac{\mu_m C_L}{K_s + C_L} - DX, \tag{47}$$

$$\frac{dC_L}{dt} = k_L a (C_L^* - C_L) - \frac{1}{Y_{O_2}} \frac{\mu_m C_L X}{K_s + C_L} - D C_L, \tag{48}$$

where X cell concentration,
μ_m maximum specific growth rate,
K_s affinity constant for oxygen,
D dilution rate, and
Y_{O_2} yield coefficient for oxygen.

For steady state conditions and for very low dissolved oxygen concentrations the cell production rate is given by:

$$XD \cong (k_L a) C_L^* Y_{O_2}. \tag{49}$$

In systems with distributed parameters (e.g. in bubble columns) $(k_L a)$ and C_L^* are position dependent. Therefore Eq. (49) is to be replaced by more complex equations [100]. However, the conclusions which can be drawn from Eq. (49) are valid for bubble columns too.

According to Eq. (49) the cell production rate can be increased.

- by application of a fermentor with high volumetric mass transfer coefficient $(k_L a)$;
- by increase of C_L^*, e.g. by means of increase of the pressure or the oxygen concentration in air;
- by application of a substrate with high yield Y_{O_2}.

As has been shown in section 7, very high volumetric mass transfer coefficients can be achieved in bubble column fermentors already by relatively low energy intake. Furthermore, it is easy to increase C_L^* by applying tall columns.

Because of the low investment and variable costs of bubble column fermentors the widening of their application is expected for non or slightly coalescing fermentation media with not too high a viscosity.

Their application for SCP production is treated in the parallel article in which more detailed economic calculations are presented [100].

Acknowledgement

The authors gratefully acknowledge the financial support of the "Bundesministerium für Forschung und Technologie" of the German Federal Government and thank the "Gesellschaft für Biotechnologische Forschung mbH" for much appreciated cooperation.

Nomenclature

(M = mass, L = length, T = time)

$A = \frac{a'}{V}$	specific surface area	L^{-1}
A'	surface area general	L^2
$a = \frac{a'}{V_L}$	specific surface area	L^{-1}
a'	gas liquid interfacial area	L^2
a^*	radius of the jet	L
$Bd^2 = \frac{\rho_L g d_B^2}{4\sigma}$	Bond number	–
$Bo = \frac{w_L H}{D_{LB}}$	Bodenstein number	–
C_L, C_L^*	concentration of O_2 in the liquid bulk phase, at the interface	ML^{-3}
$C_{LG} = C_L^*$	concentration of O_2 in the liquid in equilibrium with the gas phase	ML^{-3}
$C^* = \frac{C_L}{C_L^*}$	dimensionless concentration	–
C	constant	
D	dilution rate	T^{-1}
D_{LB}	coefficient of backmixing	L^2T^{-1}
$D_{L\,eff}$	coefficient of longitudinal dispersion	L^2T^{-1}
D_m	molecular diffusivity	L^2T^{-1}
$D_{ö}$	diameter of orifice or nozzle	L
D_s	diameter of stirrer	L
d_B	bubble diameter	L
$\bar{d}_B$	mean bubble diameter	L
d_s	Sauter mean bubble diameter	L
E	rate of energy dissipation, power input	ML^2T^{-3}
F	feed	L^3T^{-1}
$Fr = \frac{w_{SG}}{\sqrt{g d_s}}$	Froude number	–
$G = \frac{Q}{V_L}$	OTR oxygen transfer rate	$ML^{-3}T$
$Gr = \frac{d_B^2 \rho_L \Delta\rho g}{\eta_L^2}$	Grashoff number	–
g	acceleration of gravity	LT^{-2}
H	height of the bubbling layer	L
H_L	height of the bubble free layer	L
He	Henry constant	L^2T^{-2}
K_s	affinity constant for oxygen	ML^{-3}
k_L	mass transfer coefficient	LT^{-1}

k	wave numbers of disturbances	L^{-1}
L	length of the test section	L
L	characteristic length of the system	L
l	scale of turbulence	L
N	stirrer speed	T^{-1}
m	constant	
n	constant	
P	pressure	$ML^{-1}T^{-2}$
$Pr = \frac{dX}{dt}$	cell productivity	$MT^{-1}L^{-3}$
$Pe = \frac{w_L L}{D_{L\ eff}}$	Peclet number	–
p	constant	
p	local pressure	$ML^{-1}T^{-2}$
Q	gas flow rate	L^3T^{-1}
Q_{O_2}	mass flow of O_2	MT^{-1}
$Re_N = \frac{ND_s^2}{\nu_L}$	stirrer Reynolds number	–
$Sc = \frac{\nu_L}{D_m}$	Schmidt number	–
$Sh = \frac{k_L d_B}{D_m}$	Sherwood number	–
$St = k_L a \bar{t}$	Stanton number	–
t	time	T
$\bar{t}$	mean residence time of liquid	T
X	cell concentration	ML^{-3}
$X = \frac{x}{L}$	dimensionless longitudinal coordinate	–
X_0	cell concentration in-feed	ML^{-3}
x	longitudinal coordinate	L
Y_{O_2}	oxygen yield coefficient	–
V	volume of bubbling layer	L^3
V_L	volume of bubble free layer	L^3
w_G	effective gas velocity	LT^{-1}
w_L	effective liquid velocity	LT^{-1}
w_R	relative gas velocity	LT^{-1}
w_{SG}	superficial gas velocity	LT^{-1}
w_{SL}	superficial liquid velocity	LT^{-1}
$We = \frac{\tau d_B}{\sigma}$	Weber number	–
$We^2 = \frac{\rho_L w_R^2 d_B}{2\sigma}$	Weber number for freely ascending bubbles	–
w_T	terminal rising velocity of bubbles	LT^{-1}
w_L^*	local mean liquid velocity	LT^{-1}
α	growth rate of the disturbance	T^{-1}

δ_p	pore diameter of porous plate	L
E	porosity of porous plate	–
$E_G = \frac{V - V_L}{V}$	$= \frac{H - H_L}{H}$ mean relative gas holdup	–
η_L	dynamic viscosity of liquid	$MT^{-1}L^{-1}$
μ	specific growth rate	T^{-1}
$\nu_L = \frac{\eta_L}{\rho_L}$	kinematic viscosity of liquid	L^2T^{-1}
ρ_L	density of liquid	ML^{-3}
ρ_G	densitiy of gas	ML^{-3}
$\Delta\rho = \rho_L - \rho_G$		ML^{-2}
σ	surface tension	ML^{-2}
τ	turbulent shear stress	$ML^{-1}T^{-2}$

The units of the coordinates are given on the figures according to the German standard, i.e. w_{SG}/cm/s means w_{SG}[cm/s].

References

1. Atkinson, B.: Biochemical reactors. Pion Limited London, 1974.
2. Blakebrough, N.: "Industrial Fermentations", in: Biochemical and Biological Engineering Science, Vol. 1, Ed. N. Blakebrough, (Acad. Press New York) pp. 25–47, 1967.
3. Ault, R. G., Hampton, A. N., Newton, R., Roberts, R. H.: I. Inst. Brew. **75**, 261 (1969).
4. Greenshields, R. N., Smith, E. L.: The Chemical Engineer, May, 182 (1971).
5. Kloud, J., Strebacek, Z.: Chem. Techn. **24**, 688 (1972).
6. Falch, E. A., Gaden, E. L.: Biotechn., Bioeng. **11**, 927 (1967).
7. Falch, E. A., Gaden, E. L.: Biotechn., Bioeng. **12**, 465 (1970).
8. Royston, M. G.: Brit. Patent 929315.
9. Shore, D. T., Watson, E. G.: Brit. Patent 938173.
10. Shore, D. T., Royston, M. G., Watson, E. G.: Brit. Patent 959049.
11. Hall, R. D., Howard, G. A.: Brit. Patent 979491.
12. Royston, M. G.: Brit. Patent 1071428.
13. Royston, M. G.: Brit. Patent 1068414.
14. Royston, M. G.: Process Biochem. **1**, 215 (1966).
15. Klopper, W. J., Roberts, R. H., Royston, M. G.: Proc. Eur. Brew. Conv. 238 (1965).
16. Ault, R. G.: Proceeding of tenth congress, Stockholm, of European Brewery Convention (Holland: Elsevier Publ. Co.) p. 238 (1965).
17. Shore, D. T., Royston, M. G.: Chem. Eng. London, 218 May CE 99 (1968).
18. Le Francois, L., Mariller, C. C., Mejane, J. V.: France patent No. 1102200, 4 May 1955.
19. Le Francois, L.: Proceedings of the 34th International Conference of Ind. Chem. Section 14, Fermentation Industries, Sept. 22–29, Belgrade 1963.
20. Le Francois, L.: Chim. Ind. **8**, 1038 (1969).
21. Hatch, R. T.: Ph. Thesis MIT, Cambridge, Mass. 1973.
22. Hatch, R. T.: Fermentor Design, Chapter 3 in "Single Cell Protein", Ed. S. R. Tannenbaum, R. I. C. Wang, MIT Press, p. 46, 1975.
23. Iyengar, M. S., Barnach, J. N.: Brit. Chem. Eng. **13**, 684 (1968).
24. Coty, C. F., Gorring, R. L., Heiloweil, I. L., Leavitt, R. I., Srinivasan, S.: Biotechn. Bioeng. **13**, 825 (1971).

25. Prokop, A., Sobotha, M.: Insoluble Substrate and Oxygen Transport in Hydrocarbon Fermentation. Chapter 6, in "Single Cell Protein II", Ed. S. R. Tannenbaum, D. I. C. Wang, MIT Press Cambridge Mass. 1975, p. 127.
26. Gow, J. S., Littlehails, J. D., Smith, S. R. L., Walter, R. B.: SCP-Production from Methanol: Bacteria. Chapter 8 in "Single Cell Protein II", Eds. S. R. Tannenbaum, D. I. C., Wang, MIT Press 1975, p. 370.
27. Lainé, B. M., du Chaffant, J.: Gas Oil as a substrate for Single Cell Protein Production. Chapter 21 in "Single Cell Protein II", Ed. S. R. Tannenbaum, D. I. C. Wang, MIT Press 1975, p. 424.
28. Kanazana, M.: The Production of yeast from n-Paraffins, Chapter 22. in "Single Cell Protein II", Ed. S. R. Tannenbaum, D. I. C. Wang, MIT Press 1975, p. 438.
29. Cooper, P. G., Silver, R. S., Boyle, J. P.: Semi Commercial Studies of a Petroprotein Process Based on n-Paraffins, Chapter 23 in "Single Cell Protein II", Ed. S. R. Tannenbaum, D. I. C. Wang, MIT Press 1975, p. 454.
30. Calderbank, P. H., Moo-Young, M. B.: Chem. Eng. Sci. **16**, 39 (1961).
31. Calderbank, P. H.: Mass transfer in Fermentation Equipment, Chapter 5 in "Biochemical and Biological Engineering Science", Vol. 1 (1967) Ed. N. Blakebrough, Acad. Press, p. 101.
32. Oels, U., Schügerl, K., Todt, J.: Chem. Ing. Techn. **48**, 73 (1976).
33. Fan, L. T., Chen, G. K. C., Erickson, L. E.: Chem. Eng. Sci. **26**, 379 (1971).
34. Chen. G. K. C., Fan, L. T., Erickson, L. E.: Can. J. Chem. Eng. **50**, 157 (1972).
35. Deckwer, W., Popovich, M.: Chem. Ing. Techn. **45**, 984 (1973).
36. Todt, J.: Doctoral Thesis, TU Hannover 1974.
37. Østergaard, K., Michelsen, M. L.: Can. J. Chem. Eng. **47**, 107 (1969).
38. Berghmans, J.: Chem. Eng. Sci. **28**, 2005 (1973).
39. Kolmogoroff, A. N.: Compt. rend. acad. sci. U. S. S. R. **30**, 301 (1941), see also in "Turbulence" Classic papers on statistical theory. Ed. Friedlaender, S. K., Topper, L., Interscience Publ. New York, 1961, p. 159.
40. Hinze, J. O.: A. I. Ch. Journal **1**, 289 (1965).
41. Batchelor, G. K.: Proc. Camb. Phil. Soc. **47**, 359 (1951).
42 a. Nagel, O., Kürten, H., Sinn, R.: Chem. Ing. Techn. **44**, 367 and 899 (1972).
42 b. Nagel, O., Kürten, H., Hegner, B.: Chem. Ing. Techn. **45**, 913 (1973).
43 a. Levenspiel, O.: Chemical Reaction Engineering. John Wiley Sons, New York 1962.
43 b. Levenspiel, O., Bishoff, K. B.: Advances in Chemical Engineering, Vol. **4** (1963) Ed. Drew, T. B. et al., Academic Press New York, p. 95.
44. Kumar, R., Kuloor, N. R.: Advances in Chemical Engineering, Vol. **8** (Ed. Drew, T. B., et al.) Acad. Press New York 1970.
45. McCann, D. J., Prince, R. G. H.: Chem. Eng. Sci. **26**, 1505 (1971).
46. McCann, D. J., Prince, R. G. H.: Chem. Eng. Sci. **24**, 801 (1969).
47. Jameson, G. I., Kupferberg, A.: Chem. Eng. Sci. **22**, 1053 (1967).
48. Kupferberg, A., Jameson, G. I.: Trans. Instn. Chem. Engrs. **47**, T 241 (1969).
49. Marmur, A., Rubin, E.: Chem. Eng. Sci. **31**, 453 (1976).
50. Oels, U., Lücke, J., Schügerl, K.: Chem. Ing. Techn. **49**, 59 (1977).
51. Gestrich, W., Rähse, W.: Chem. Ing. Techn. **47**, 8 (1975).
52. Gestrich, W., Krauss, W.: Chem. Ing. Techn. **47**, 361 (1975).
53. Gestrich, W., Eisenwein, H., Kraus, W.: Chem. Ing. Techn. **48**, 399 (1976).
54. Eissa, S. H., Schügerl, K.: Chem. Eng Sci **30**, 1251 (1975).
55. Sahm, H., Wagner, F.: Arch. Microbiol. **84**, 29 (1972).
56. Linek, V., Mayrhoferova: Chem. Eng. Sci. **25**, 787 (1970).
57. Oels, U.: Doctoral thesis. Technical University Hannover 1975.
58. Todtenhaupt, E. K.: Chem. Ing. Techn. **43**, 336 (1971).
59. Pilhofer, T.: Chem. Ing. Techn. **46**, 913 (1974).
60. Buchholz, R.: Diplomarbeit. Technical University Hannover 1976.
61. Himmelblau, D. M.: Process Analysis by Statistical Methods, John Wiley New York (1970).
62. Marquardt, D. W.: Chem. Eng. Progr. **55** (6) 65 (1959).

63. Deckwer, W. D., Burckhart, R., Zoll, G.: Chem. Eng. Sci. **29**, 2177 (1974).
64. Langemann, H.: Chem. Ztg./Chem. Apparatur **92**, 845 (1968).
65. Koide, K., Kato, K., Tanaka, Y., Kubota, H.: J. Chem. Eng. of Japan **1**, 51 (1968).
66. Akita, K., Yoshida, F.: Ind. Eng. Chem. Process Des. Development **12**, 76 (1973).
67. Akita, K., Yoshida, F.: Ind. Eng. Chem. Process Des. Development **13**, 84 (1974).
68. Hayashi, T., Koide, K., Sato, T.: J. Chem. Eng. of Japan, **8**, 16 (1975).
69. Calderbank, P. H., Moo-Young, M. B., Bibby, R.: Third European Symposium on "Chemical Reaction Engineering" 1964. Pergamon Press.
70. Lücke, J., Oels, U., Schügerl, K.: Chem. Ing. Techn. **48**, 573 (1976).
71. Hughmark, G. A., Ind. Eng. Chem. Proc. Des. Development **6**, 218 (1967).
72. Burkel, W.: Doctoral Thesis TU München (1974).
73. Marucci, G.: Chem. Eng. Sci. **24**, 975 (1969).
74. Nagel, O., Kürten, H.: Chem. Ing. Techn. **48**, 513 (1976).
75. Bowonder, B., Kumar, R.: Chem. Eng. Sci. **25**, 25 (1970).
76. Nagel, O.: Personal communication.
77. Reith, T.: Doctoral thesis, TH Delft (1968).
78. Chang, C. L.: Doctoral thesis, TU Berlin (1968).
79. Forth, K.: Doctoral thesis, TU Berlin (1966).
80. Levenspiel, O., Godfrey, J. H.: Chem. Eng. Sci. **29**, 1723 (1974).
81. Katinger, H. W. D.: in 3. Symposium technische Mikrobiologie, Berlin 1973 p. 95.
82. Brauer, H.: Chem. Ing. Techn. **45**, 1099 (1973).
83. Davies, J. D.: Mass Transfer and Interfacial Phenomena in "Advances in Chemical Engineering" Vol. 4, Ed. Drew, T. B., Hoopes, Jr., J. W., Vermeulen, T.; Acad. Press New York, 1963, p. 1.
84. Brauer, H.: Stoffaustausch einschließlich chemischer Reaktionen. Verlag Sauerländer, Aarau und Frankfurt (1971).
85. Higbie, R.: Trans. Am. Inst. Chem. Engn. **35**, 365 (1935).
86. Reuss, M.: Doctoral thesis, TU Berlin (1970).
87. Reith, T., Renken, S., Israel, B. A.: Chem. Eng. Sci. **25**, 619 (1958).
88. Langemann, H., Taubert, C.: Verfahrenstechnik **2**, 417 (1968).
89. Argo, W. B., Cova, D. R.: Ind. Eng. Chem. Proc. Des Development **4**, 352 (1965).
90. Towell, G. D., Ackermann, G. H.: Fifth European, Sec. Intern. Symp. on Chem. React. Engng. (1972).
91. Hinze, J. O.: Turbulence, 2nd Edition, McGraw Hill Co., New York 1975.
92. Serizawa, A., Kataoka, I., Michiyoshi, I.: Int. J. Multiphase Flow **2**, 221 (1975). **2**, 235 (1975), **2**, 247 (1975).
93. Sato, Y., Sekoguchi, K.: Int. J. Multiphase Flow **2**, 79 (1975).
94. Davis, R. E., Acrivos, A.: Chem. Eng. Sci. **21**, 681 (1966).
95. Saville, D. A.: The Chem. Engng. Journal **5**, 251 (1973).
96. Hadamard, J.: Compt. Rend. Acad. Sci. **152**, 1735 (1911).
97. Rybczynski, W.: Polska Akademija Umiejetnosci, Krakow, Wydz. Mat. Przetron, Series **A 403**, 40 (1911).
98. Offenlegungsschrift Federal Republic of Germany 1557018. 9. 4. 1966.
99. Zlokarnik, M.: Article in Vol. 8.
100. Schügerl, K.; Lücke, J.; Lehmann, J., Wagner, F.: Article in Vol. 8.
101. Taylor, G. I.: Proc. Roy. Soc. **A 219**, 186 (1953); **A 225**, 446, 473 (1954).
102. Schügerl, K.: Chem. Eng. Sci. **22**, 793 (1967). Schügerl, K.: Experimental comparison of mixing processes in two- and three-phase fluidized beds, in "Proceedings of the internat. Symp. on Fluidization", Ed. A. A. H. Drinkenburg, Netherlands Univ. Press, Amsterdam, 1967, p. 782.
103. Ebner, H.: 3. Symposium Technische Mikrobiologie Berlin 1973, Inst. f. Gärungsgewerbe und Biotechnologie, Hrsg. H. Dellweg, 1973, p. 71.
104. Hirner, W.: Doctoral thesis, University Stuttgart 1974.
105. Hassan, I. T. M., Robinson, C. W.: Fifth International Fermentation Symposium Berlin 1976, Ed. H. Dellweg, p. 58.

106. Einsele, A.: Fifth International Fermentation Symposium Berlin 1976, Ed. H. Dellweg, p. 69.
107. Fuchs, R., Ryn, D. D. Y., Humphrey, A. E.: Ind. Eng. Chem. Proc. Des. Develop. **10**, 190, (1971).
108. Brauer, H.: Grundlagen der Einphasen- und Mehrphasenströmungen, Sauerländer, Aarau/Frankfurt/M. 1971.
109. Ruff, K.: Chem. Ing. Techn. **46**, 769 (1974).
110. Meister, B., Scheele, G. F.: A. I. Ch. E. J **13**, 682 (1967).
111. Rayleigh, Lord: Phil. Mag. **34**, 177 (1892).
112. Tyler, E.: Phil. Mag. **16**, 504 (1933).

Description and Operation of a Large-Scale, Mammalian Cell, Suspension Culture Facility

R. T. Acton and J. D. Lynn
Department of Microbiology, University of Alabama in Birmingham,
Birmingham, AL 35294, USA

Contents

Introduction

The establishment of improved methods for the growth of various cell types *in vitro* has opened up many new areas of investigation. There is an ever increasing demand for large quantities of subcellular organelles, normal cellular products and viruses which are being utilized in academic as well as industrial laboratories. This demand often becomes quite acute when viral infectious disease epidemics are predicted and mass quantities of virus for vaccine production are needed for immunization programs. It is therefore not surprising that the wide use of cultured cells has stimulated the development of techniques and systems for large-scale production.

Obviously, the type of cell(s) one wishes to propagate dictates to a large degree the design of any large-scale culture system. The system operative in our laboratory was designed for the growth of cells in submerged or suspension culture. Although BHK 21 cells have been grown by this method on an industrial scale in volumes up to 2000 l, there have been relatively few reports of other large-scale submerged culture systems [1–16]. Table 1 is a compilation of systems, excluding our own, 10 l and larger that have been utilized for the propagation of mammalian cells. In some of these systems attempts were made to control environmental parameters such as temperature, pH and dissolved oxygen (DO) or oxidation-reduction potential (ORP). To date no one has reported a detailed study whereby environmental parameters, nutritional requirements, cell growth and the amount of cell product has been observed and correlated in a large-scale system. Moreover, there are only reports of 7 different types of cells grown on a large-scale. This paucity of information led our group to systematically investigate the

Table 1. A comparison of large-scale systems for propagating mammalian cells in suspension culture

System	Agitation	Volume (l)	Cell Type	Reference
Round-bottom flask	Impeller	10	Human cervical carcinoma (Hela)	Holmström (1964)
New Brunswick fermentor	Impeller	14	Mouse bone marrow (JLS-V9)	Hodge *et al.* (1974)
New Brunswick fermentor	Impeller	14	Human lymphoblastoid	Mizrahi *et al.* (1971)
Belco spinner flask	Impeller	16	Burkitt Lymphoma (P_3HR-1)	Klein *et al.* (1976)
Glass carboy	Vibromixer	18	P_3HR-1	Klein *et al.* (1976).
Custom fermenter	Impeller	20	Hela Mouse fibroblast (L) Embryonic rabbit kidney (KD)	Zeigler *et al.* (1958)
Pyrex aspirator	Magnetic stirring bar	20	Embryonic rabbit kidney (ERK)	Cooper *et al.* (1959)
Custom fermentor	Impeller	30	Baby hamster kidney (BHK-21)	Telling and Elsworth (1965); Telling *et al.* (1967); Radlett *et al.* (1971); Radlett *et al.* (1972)
Custom fermentor	Vibromixer	30	BHK-21	Fontanges *et al.* (1971)
Custom fermentor	Impeller	40	Mouse fibroblast (LDR)	Klein *et al.* (1971)
Custom fermentor	Impeller	40	LDR	Wiles & Smith (1969)
Custom fermentor	Impeller	100	HeLa	Holmström (1964)
Custom fermentor	Vibromixer	200	Human Lymphoblastoid	Moore *et al.* (1968)
Custom fermentor	Vibromixer	300	BHK-21	Girard *et al.* (1973)
Custom fermentor	Impeller	2000	BHK-21	Telling and Radlett (1970)

growth parameters of cells in small-scale culture before designing a large-scale facility. As will be described in this review we have defined conditions for optimal growth of murine lymphoblastoid cell lines from a 1 l spinner flask up to a 200 l fermentor-type vessel. The design of equipment necessary to effect the growth of cells and product acquisition on a large-scale will also be discussed.

1. Determination of Cell Growth Parameters

In most small-scale suspension culture operations cells are propagated under semi-continuous culture conditions. This mode of operation represents essentially a "feast-famine" situation whereby cells are inoculated into fresh medium at a relatively low density. A gas mixture of CO_2 and air is then introduced. Growth continues until a build-up of toxic metabolic products or a depletion of nutrients slows or stops cell growth entirely. Obviously, if one is interested in efficiently producing cells in large quantities, the factors limiting cell growth must be understood and ultimately controlled. An examination of the events which occur in a theoretical, semi-continuous suspension culture of mammalian cells is enlightening and indicates which parameters must be controlled for optimal growth. The growth curve of mammalian cells in suspension culture is a semi-logarithmic plot of cell density (cells/ml of culture fluid) as a function of time (h). The resulting semi-logarithmic curve will be of the sigmoidal variety. Figure 1 is a theoretical mammalian cell growth curve. The first segment of the curve, 1, will be referred to as the "lag phase" of growth. Cells in the lag phase undergo little or no division. The slope of the growth curve for the lag phase is approximately 0 (slope and the second derivative are determined using a superimposed Cartesian ordinate axis graduated as the abscissa on the semi-logarithmic growth plot). When the second derivative of the curve changes from 0 to a positive value, the cells have entered the terminal lag phase of growth, segment 2. At this stage, the cells have begun dividing. In this phase of growth the cell division rate is constantly increasing. That is, the second derivative of this portion of the curve is a positive value. When the second derivative of the curve changes from a positive value to 0, the curve has a linear slope which has an average value of between about 1 and 100, segment 3. At this point the cells have entered what is referred to as the "exponential phase" of growth. It is during this stage that the cells are undergoing division at a constant maximum rate. The cells continue growing at this rate until the cell concentration becomes so large or nutrients in the medium are depleted that the cells enter the "early stationary phase" of growth, segment 4. In this phase of growth, the cells' division rate decreases. The onset of the early stationary phase occurs at that point where the second derivative of the curve changes from 0 to a negative value. If the cells are allowed to continue, they will enter the "stationary phase" of growth, segment 5. At this stage of growth, the cells' mortality rate increases and net production decreases. Accordingly, the slope of this segment of the curve is approximately 0. The onset of this stage of growth occurs at the point where the second derivative changes from a negative value to approximately 0. If the cells are allowed to remain in the stationary phase for too long, they cease to divide and a net loss of cells will result.

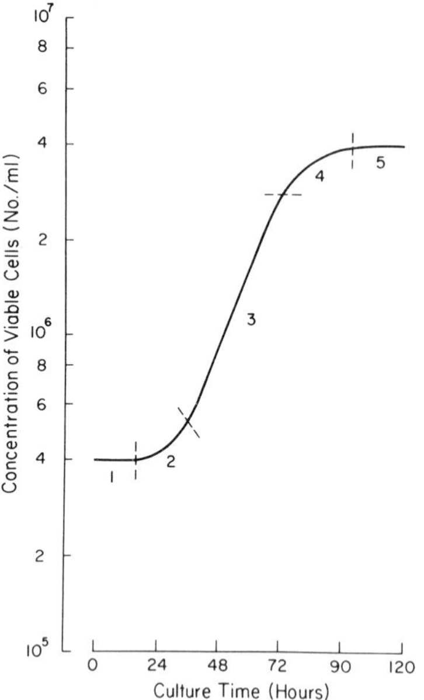

Fig. 1. Idealized growth curve

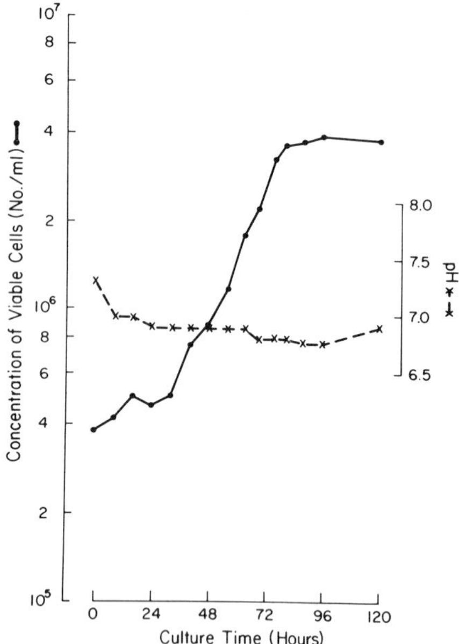

Fig. 2. Growth of S49.1 in a 14 l New Brunswick fermentor utilizing RPMI-1640 supplemented with 10% horse serum (HS) as the growth medium. CO_2 was continuously added to the vessel by overlay until the pH reached 6.8 at which time a mixture of 5% CO_2 + air was continuously added by overlay. The gas flow rate was 85.7 ml/min. Impeller speed was 300 rpm

That mammalian cells grown in suspension culture do indeed perform as just described can be illustrated by referring to Fig. 2. This figure defines the growth properties of the murine lymphoblastoid cell line S49.1 in a modified 14 l New Brunswick fermentor [17–19]. In this particular example S49.1 was introduced into the fermentor at a concentration of approximately 4×10^5 cells/ml. For approximately 32 h after inoculation the cells grew very little and were in the lag phase of growth. Soon after this interval, the cells entered the exponential phase of growth where doubling times of 8, 14, and 30 h respectively were observed until the cells reached a maximum of 4×10^6 cells/ml. As soon as the cells reached this maximum density, growth ceased and 24 h later a small reduction in cell number was observed. Thus, the cells had entered the stationary phase of the growth curve. It was observed that cells taken from the stationary phase and reinoculated into fresh medium always failed to divide until 24–48 h had elapsed. However, if cells were reintroduced into fresh medium while in the exponential or very early stationary phase of growth, the cells continued to divide at a rapid rate and did not enter a lag phase. One could speculate that once cells cease to divide due to a depletion of nutrients or a buildup of toxic by-products of metabolism, growth and division are in someway suppressed or terminated. Therefore, when cells are provided fresh medium, it takes a certain period of time for this machinery to be turned on again. Whatever the reason, it was important for our operation to have discovered this feature of cell growth which provided quidelines for culturing cells in semi-continuous modes. A set of conditions have now been established which prove useful for the growth of murine lymphoblastoid cell lines in large-scale culture.

When culturing mammalian cells according to this scheme, the cells are repetitively cultured from a cell concentration approximately corresponding to the terminal lag phase of growth to a cell concentration corresponding to the early stationary phase of growth. A sufficient amount of medium and cells are then harvested to allow dilution of the cells in the remaining suspension with fresh growth medium to a cell concentration corresponding to approximately the terminal lag phase of growth. The fresh growth medium is then added in a quantity sufficient to achieve this concentration.

Preferably, the cell concentration approximately corresponding to the terminal lag phase is in the range of from where the second derivative of the growth curve is at its maximum positive value up to 20% greater than where the second derivative changes from a positive value to approximately 0. More preferably, the cell concentration is in the range of from 10% less to 10% greater than where the second derivative changes from a positive value to 0.

The preferred cell concentrations corresponding to the early stationary phase of growth range from 20% less than the point where the second derivative changes from 0 to a negative value to where the slope of the growth curve has decreased to 30% of the value of the average slope in the region of exponential growth. More preferably, the cell concentration is in the range of from 0 to a negative value to where the slope of the growth curve has decreased to 40% of the value of the average slope of the growth curve in the exponential region of growth.

In practice one may establish a growth curve and continuous culture conditions by the following method:

1) begin culture at increasing initial densities starting at some concentration, e.g., 10^5 cells/ml,

2) when the lowest initial concentration is found that results in a very short or no lag phase, the optimal inoculation point has been reached,

3) note slope of exponential phase and beginning of stationary phase during latter cultures,

4) innoculate at newly determined point,

5) monitor slope of exponential phase,

6) when it decreases to 50% of its average value, remove a volume of fluid equal to the amount of medium that would be required to reduce the cell density to the optimal value found earlier,

7) efficient semi-continuous culture may now begin; a slight improvement may be realized by varying the two experimentally determined end points.

By following these guidelines we have been able to culture murine lymphoblastoid cell lines for long periods of time achieving a remarkable degree of reproducibility. As can be seen in Fig. 3 the BW5147 murine lymphoblastoid cell line [20] has been cultured in the 14 l fermentor for 36 continuous days [21]. Maximum cell densities varied from

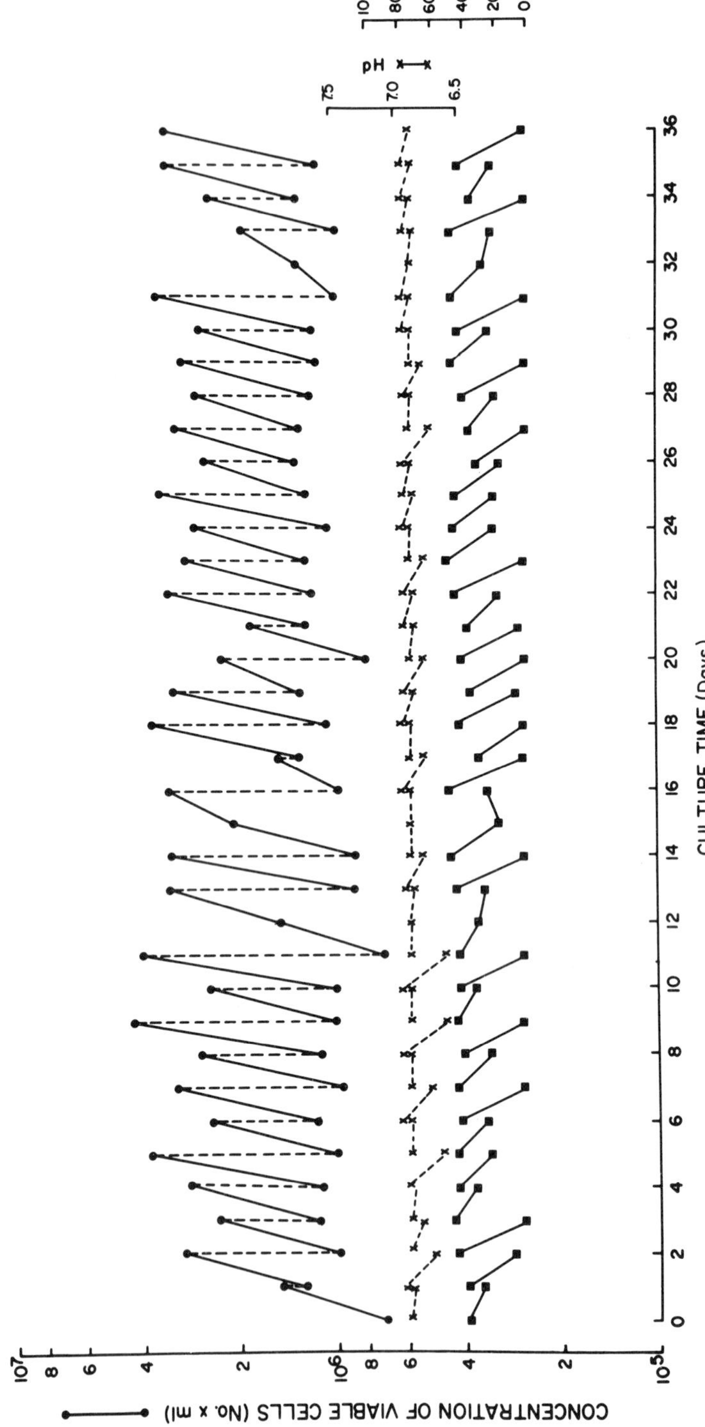

Fig. 3. The growth of BW5147 under semi-continuous conditions in a 14 l New Brunswick fermentor utilizing RPMI-1640 supplemented with 0.3 g/l asparagine, 2.5 g/l glucose and 10% horse serum as the growth medium. CO_2 was continuously added to the vessel by sparge as well as overlay until the pH reached the set-point (6.8 initially) after which air was added continuously by sparge. The gas flow rate was 85.7 ml/min and impeller speed was 300 rpm

$3-4 \times 10^6$ cells/ml during this culture period and the doubling time ranges from 12 to 22 h. The fact that cells could be propagated under these conditions for 36 continuous days without microbial contamination, alteration of growth properties or change in cell surface phenotype speaks for the feasibility of our approach. Moreover, a total of 5×10^{11} viable cells were harvested during this period.

These types of experiments on a small-scale influenced our thinking considerably with regard to the design and subsequent operation of our large-scale facility. As will be described later in this review the parameters observed for cell growth in the 14 l fermentor are comparable to those in the large-scale system.

2. Design Considerations

This section is dedicated to the physical description of the equipment and facilities used to culture mammalian cells on a large-scale in suspension. Apparatus required to grow cells in small to intermediate quantities is described briefly here and more extensively heretofore [22] by the authors. Three subsections will follow describing the culture instrumentation, its support equipment and the facility in which it is housed.

a) Primary Equipment

In general the culture apparatus is composed of stainless steel vessels, control cabinets and separation equipment as can be seen in Fig. 4. One can get a better feel for the facility by comparing the layout to the photograph taken from the inoculum preparation laboratory doorway (Fig. 5).

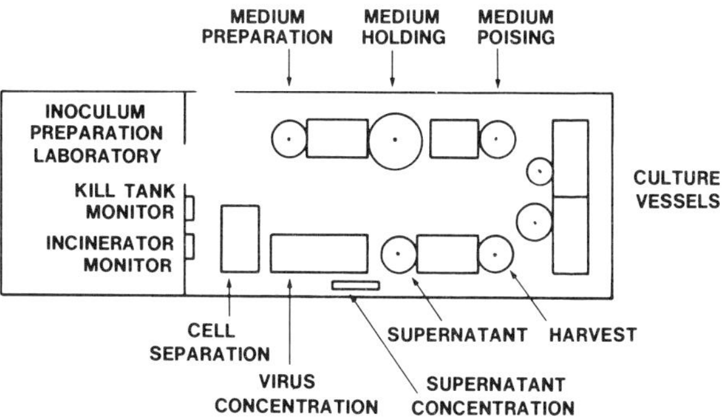

Fig. 4. Primary equipment layout

The process of cell growth in large volumes begins with medium preparation. Medium is prepared by adding concentrate and serum to high purity water which is contained in a 650 l non-pressurized vessel. The medium preparation vessel has an agitation system to insure rapid and total solution of all the components of the medium. This system is composed of a magnetically driven, offset, marine impeller located in the bottom of the vessel [23]. This configuration obviates the need for seals or diaphragms in that there is

Fig. 5. Photograph of primary equipment

no penetration of the vessel. Also, by virtue of its eccentricity, there is no need for baffles or other turbulence causing devices. The impeller is driven by a gearhead motor which has a permanent magnet attached to it similar to the one inside the vessel. The speed of the agitator can be monitored and controlled from a control panel adjacent to the vessel. At this stage the pH of the medium may be adjusted by the addition of acid or base which is controlled by pH monitor, controller and recorder located at the top of the control panel. In our approach, final adjustment of pH is made in the poising vessel which will be discussed presently.

After the medium has been prepared, it is then ready for sterilization. The medium passes through a quarter-turn, stainless steel ball valve with Teflon packing by virtue of a centrifugal pump through a pair of steam presterilized (in place) Pall Trinity cartridge filters and into the medium holding vessel. The two series cartridge filters (bacterial prefilter, AB2AA8P and a AB2AR8P 0.22 μm) are in parallel with a similar pair of filters. Both types of filters have 10 square feet of surface area. In the event one set becomes blocked, an appropriate valve manipulation can direct flow through the alternate set of filters into the holding tank. The holding vessel is a 625 l stainless steel pressure vessel which has a jacket around its circumference to allow cooling fluid to circulate in order to keep the temperature of the vessel depressed. This vessel is also insulated on the outside to avoid condensation of water vapor. This vessel along with the remaining vessels of the system is jacketed, of the pressure type and steam sterilizable in place. Medium can be held in this vessel for extended periods of time and removed as needed. It is normally held at 2° C with the agitator operating to insure homogeneous temperature and inhibition of precipitation. There is also a control panel adjacent to the vessel. A set of valves are located at the lower part of the control panel which allow regulation of air pressure in the vessel, circulation of coolant and steam sterilization. Above these valves in the same panel are the temperature controlling, monitoring and recording instrumentation along with that for agitation.

The next step in the process is to prepare the medium immediately before adding it to the cell culture. This procedure is called "medium poising." During the poising procedure, the temperature of the medium is increased to that of the culture temperature, 37 °C. The reduction-oxidation potential can be adjusted to a value compatible with the particular cell line of interest. The pH is also adjusted to its appropriate value by sparging CO_2 through the liquid. The temperature, pH and E_h are monitored, controlled and recorded at this point by the gear located in a relay rack as described for the other vessels. There is also a set of valves (similar for all vessels) at the bottom of the relay rack which are used for temperature control, gas regulation and steam sterilization. The poising vessel is not insulated as is the medium holding vessel. Since the temperature of this vessel remains above room temperature, there is no tendency for condensate formation. Transfer of the medium from the holding vessel to the poising vessel is accomplished through a teflon line coated with braided stainless steel. The line is connected to the base of the holding vessel by virtue of a Swagelok quick-connect and then to the poising vessel with another quick-connect. After the attachment is made, steam can be circulated through the flexible hose to render it sterile. After the sterilization cycle is completed, a valve at either end of the line can be opened. A pressure differential established between the two vessels allows fluid flow from the holding to the poising vessel. A liquid level indicator is located on both vessels to give indication of the amount of fluid transferred. All transfers between vessels are made in this manner.

After the medium has been poised, cell culture is now ready to begin. The medium is transferred to the culture vessel in a similar manner as the previous description and then a volume of cells is added that has been prepared in the modified 14 l fermentor. The medium and the inoculum are added to a 90 l (total volume) culture vessel. The 90 l vessel has a cell volume working capacity of approximately 70 l. Various parameters can be controlled, monitored and recorded during cell growth.

Adjacent to the vessel is a control panel, where pH, dissolved oxygen, temperature, agitation rate, turbidity and CO_2 output are monitored. Another factor is encountered at this point: we now have cells growing which may be budding viruses. We must be conscious of biohazard considerations and so, from this point on each vessel is tied to an incineration system for all gases emitted and to a kill tank system for liquids drained. These two systems will be discussed later.

After the cells in the 70 l vessel have grown to their late log phase density, they are transferred to the larger of the two culture vessels which is a 250 l vessel or to the harvest vessel or to a cell separator. For the sake of description, we will say that cells are to be directed to the larger culture vessel into which a volume of poised medium has already been added. A mixture of the inoculum and the poised medium should result in a density in accordance with the lower log phase density of that cell line. Once the 250 l vessel, (200 l working volume) is ready for harvest, cells are directed to the harvest vessel or directly into the separation system. The harvest vessel is similar to the holding tank, the only difference being the biohazard consideration. After a number of harvests, one may be ready to proceed with cell processing. This may be done by transferring from harvest tank through the flexible line to the cell separation device which is a Sharples Laboratory centrifuge enclosed in an isolator to contain hazardous aerosols. The operator works with rubber gloves in the isolator and after the process is over, sprays the interior with formaldehyde to guard against contamination. The isolator is attached to the aforementioned incinerator system.

The supernatant is directed either into a supernatant vessel for continued processing or to the kill tanks. If one is interested in products other than the cells themselves, such as viruses or macromolecules in the supernatant, additional processes are available. One may run the supernatant through a concentration system which has been developed by the Amicon Corporation. With the appropriate hollow fiber cartridges in this system, viruses can be concentrated into small volume which can then be put on the preparative ultracentrifuge, continuous-flow rotor for final concentration. This instrument is contained to eliminate contact with potential biohazard material. By changing the Amicon fiber cartridges to ones of lower molecular weight cut-off one may separate various macromolecular species found in the supernatant.

Adjacent to the cell separation device is a kill tank monitor and incinerator monitor which are for systems located in other areas.

b) Ancillary Equipment

The primary equipment which was described in the previous section requires a number of utilities for its operation. These ancillary service items are located on the floor just below the primary facility and also on the roof above. Figure 6 is a schematic representation of the ancillary equipment located on the floor below. An electric steam generator produces all the steam required to sterilize the vessels on the floor above and also the filter system and transfer lines. The large unit behind the steam generator is a refrigeration plant. This system provides chilled ethylene glycol/water mixture for the jacketed vessels on the floor above and also is used to regulate the temperature of the culture and poising vessels. The large circular vessel is a reverse osmosis water reservoir. Water is fed into this system from a unit mounted behind it on a duct chase. The water in the reservoir is continually circulated by a pump located beneath it. Figure 7 is a schematic representation of the water system. Normal tap water is fed into the system

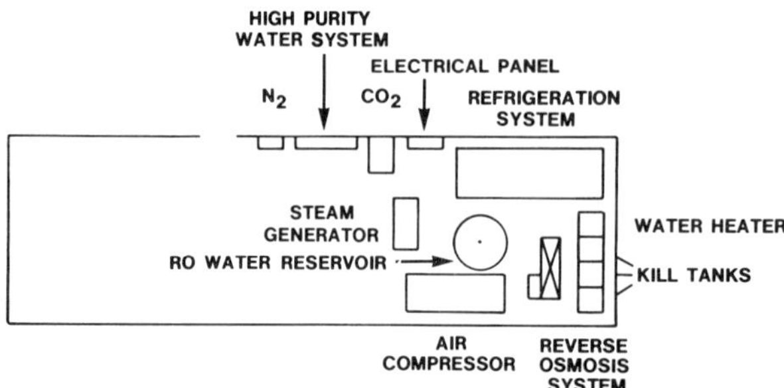

Fig. 6. Ancillary equipment layout

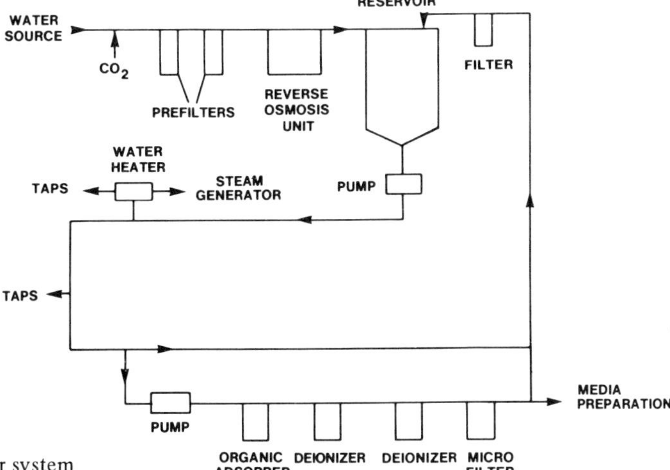

Fig. 7. Schematic of water system

along with small amounts of CO_2 gas. This mixture goes through two pre-filters mainly to remove particulate matter and then through the reverse osmosis (RO) unit. The CO_2 addition to the water increases the life of the cartridges in the reverse osmosis unit. From this unit the water goes into a 250 gal. polypropylene reservoir. A valve and pump is located at the bottom of the reservoir to circulate the water within the loop. The water in the RO loop goes to a water heater which in turn directs the water upstairs to hot water taps or to the steam generator. Also, water in the loop goes to cold water taps upstairs. The hot and cold RO water is used for washing down the vessels. This water is used to produce high purity water that is used in the preparation of medium. As one can see, there is a loop attached to the RO loop which has a pump to bring the water out of the larger loop and send it through an organic absorber, two deionizer cartridges and a micro-filter. When the water leaves these four cartridges it is very pure, of high resistivity (18 MΩ) and sterile: the properties desired for growth medium. Just before the RO water is dumped back into the reservoir there is a filter to provide resistance such that this open-type system will not drain. Not shown on the schematic are a liquid level controller in the reservoir and also a vacuum switch to shut off the pump if the reservoir goes dry. Obviously a number of check valves are required in a system of this nature that are also not shown for the sake of clarity. On either side of the high purity water system are located gas cylinder manifolds. On the left side is nitrogen which is used to regulate the reduction-oxidation potential of the medium during poising and on the right is the CO_2 manifold system which is used for pH control. An air compressor is shown which provides air for aeration during culturing of cells, to transfer liquid from one vessel to another, and to operate various pneumatic control valves. Located to the right of the compressor are four industrial water heaters. One is used to provide hot water to the steam generator and to the hose bibs on the floor above. The other three have been converted to be used as kill tanks to be described later.

On the roof above the cell culture facility are located a few items of equipment required for the operation of the system. Figure 8 is a schematic representation of the roof plan. Located on the left side of the schematic is a large exhaust fan. This fan is used to exhaust air from the two biological safety cabinets located in the facility. Located in the center is a HEPA-filtered air handler which is exclusively for the culture area. In the upper right corner of the figure is located a fan and an incinerator which is part of the biohazard containment system (Fig. 9).

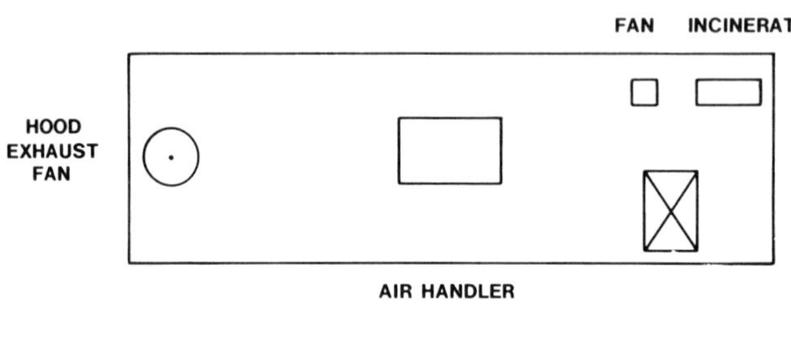

Fig. 8. Ancillary equipment layout (roof)

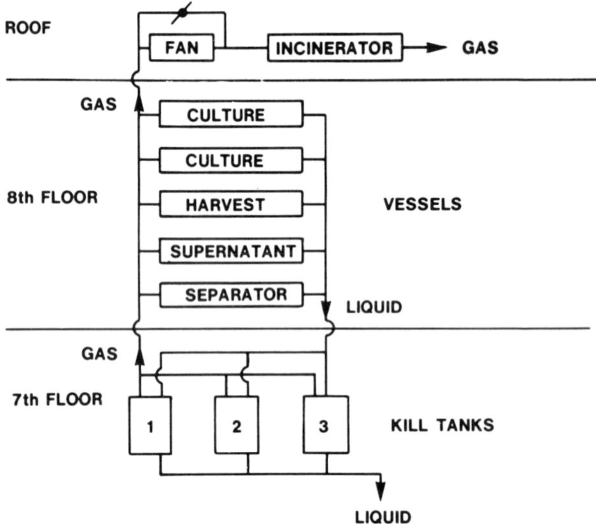

Fig. 9. Schematic of biohazard containment system

The containment system is separated into components on the seventh and eight floors and the roof. This system is capable of handling all fluids and gases emitted from the culture system. The liquids from the two culture vessels, the harvest and supernatant vessels and the cell separator are directed to the kill tanks on the floor below unless the fluids are traveling among the eighth floor vessels. The vessels on the seventh floor are industrial water heaters which have been converted to be used as kill tanks. Fluids can be transferred from one kill tank to another but they may not be released to the drain automatically, this must be done with a hand valve to avoid dumping contaminated fluid due to an electrical malfunction. A monitor in the Cell Culture room indicates the status of each tank, that is, fluid contained and length of residence. The gases in the system, that is the gases in the kill tanks and the vessels in the Cell Culture area go to the incinerator by virtue of a fan located on the roof. This incinerator has three heating elements. The status of these is monitored in the Cell Culture area. This system is designed such that if one element fails, the incinerator will still be efficient in terms of heat-killing of pathogens. Above the fan is located a balancing loop such that no matter what the conditions are the vessels on the floors below, a constant flow of gas will pass through the incinerator. If the flow drops below a certain rate, the elements will melt, and if the flow rate is too fast, there is not enough residence time for pathogenic agents to be killed. There are a number of check valves and other components in this system which have been omitted for the sake of clarity.

c) Architectural Considerations

The primary concern in housing the facility was centered around potential biohazards associated with growing large quantities of cells and processing them. The renovation was effected such that the area conforms to the guidelines as described by the National Cancer Institute for work with moderate risk oncogenic viruses. The facility also conforms to the P3 requirements of the Recombinant DNA Research Guidelines of the National Institutes of Health. The laboratory is separated from areas which are open to the general public. The surfaces of walls, floors, bench tops and ceilings are easily cleanable. The ventilation system is balanced to provide for an inflow of supply air from the access corridor into the laboratory. Exhaust air is not recirculated.

3. Facility Operation

a) Serum Preparation

It became apparent at the outset that it would be impractical to purchase sterile serum that had been screened for adventitious agents to use in the large-scale culture facility. Most supply houses provide serum in 500 ml bottles only which created the problem of introducing up to 50 l of such material into the medium holding vessel under sterile conditions. We, therefore, decided to purchase high quality serum collected under sterile conditions and stored frozen which would be processed within the facility. This "raw" serum is processed by slow thawing at 5 °C after which it is introduced into the medium poising vessel. The temperature is raised to 56 °C where it is held for 30 min under constant agitation at 100 rpm. This treatment inactivates most adventituous agents such as virus and mycoplasms which are difficult to remove by filtration. The temperature of the serum is then depressed to 5 °C after which it is rendered sterile by use of equipment designed by the Millipore Corporation. This unit consists of a 31″ AP-20 Lifegard depth filter cartridge, a 31″ AP-15 Lifegard depth filter catridge, a 10″ CW-06 Milligard prefilter cartridge, a 10″ CW-03 Milligard prefilter cartridge, a 31″ MF Millitube final filter cartridge of 0.22 μm and a 293 mm diameter membrane final disc filter of 0.22 μm. With this apparatus raw serum can be rendered sterile in 60–100 l batches depending upon lots in about 2 h. The serum is filtered directly into a series of sterile 20 l stainless steel vessels, whose configuration has been previously described [22] and held at 2 °C until use.

b) Complete Medium Preparation

Medium is prepared by adding commercially available powdered medium concentrate to high purity water in the 650 l media preparation vessel. For the cell lines that will be discussed in this review the standard formula RPMI-1640 is utilized to which 2.5 g/l glucose, 0.3 g/l asparagine and 2.0 g/l sodium bicarbonate is added. The mixture is stirred until all components have dissolved after which the desired amount of processed serum is added. After the complete medium has been prepared, it is then rendered sterile as previously discussed. We have had good success in holding batched complete medium for as long as 60 days without loss of growth capabilities or becoming contaminated with microorganisms.

Before the medium can be utilized for propagating cells it must be "poised". The pH of the medium is usually around 8.0 when introduced into the poising vessel. It is then brought to a value of 7.0 and a temperature of 37 °C before transfer to the culture vessels.

c) Cell Inoculum

The "seed" culture is prepared after recovering the cell line of interest from the frozen state by vigorously shaking ampoules in a 37 °C water bath until thawed and then diluting with complete medium so that the dimethylsulfoxide concentration is less than 0.1%. The cells are washed by centrifugation, resuspended in complete medium and inoculated into a 50 ml spinner flask. The cells are handled as described in Section 1 of this review [18] until a volume of 2 l is reached with a cell density of $1-4 \times 10^6$ cells/ml. This culture is then introduced into a 14 l New Brunswick bench-top bacterial fermentor equipped with automatic pH and dissolved oxygen control and monitor systems that have been modified to make them more amenable for the growth of mammalian cells [18]. The fermentor was fitted with marine impellers in order to minimize damage to the cells during culture. The liquid addition pumps supplied on the original instrument which were coupled to the pH controller were replaced with solenoid valves so that pH could be controlled with CO_2 and air overlay. All inlet and outlet ports were fitted with Swagelok quick-connect couplings. The procedures and equipment utilized in medium addition and cell culture harvest in conjunction with the small fermentor has previously been reported [18]. We have found that the 14 l fermentor gives an excellent indication of the types of problems to be encountered in the larger vessels as well as provides "seed" cultures for inoculation into our 70 l vessel.

As the cells continue to grow in the 14 l fermentor, fresh medium is periodically added on a semi-continuous basis until the volume reaches 12 l and the cells have grown to their maximum density without entering the stationary phase of growth. This 12 l "seed" culture is then transferred to the 70 l culture vessel. The transfer is facilitated by the fact that both vessels have quick-connect couplings which reduce the possibility of contamination. Table 2 summarizes a typical scale-up from the frozen state to the production stage of the BW-5147 cell line. Utilizing BW-5147 which usually undergoes 1–2 cell divisions in a 24 h period, it takes approximately 13 days from the time cells are retrieved from the frozen state until they can be harvested from the 70 l and 200 l culture vessels. This schedule for the generation of the seed culture can only be attained if one routinely adds fresh medium to the cells before they reach the stationary phase. In order to maintain these cells in a logarithmic stage of growth they often have to be cut twice per 24 h period. Otherwise, the cells will fail to double as rapidly or they may enter the lag phase of growth where they often die. The schedule of scale-up shown is achieved only if close attention is paid to these rules of culture.

Table 2. Generation of inoculum for production of BW-5147 cells in a 70 l and 200 l culture vessel

Day	Vessel size (1)	Inoculation 1 @ cells/ml	Medium added (1)	Procedure
0	0.050	0.003 @ 1×10^7	0.03	Frozen inoculum added to medium in 50 ml spinner flask
1	0.10	0.030 @ 1.6×10^6	0.03	Flask from day 0 added to medium in 100 ml spinner flask
2	0.25	0.060 @ 1.4×10^6	0.06	Flask from day 1 added to medium in 250 ml spinner flask
3	1	0.120 @ 2.1×10^6	0.20	Flask from day 2 added to medium in 1 l spinner flask
4	1	0.325 @ 1.8×10^6	0.32	Added medium to 1 l spinner flask from day 3
5	4	0.650 @ 2.3×10^6	0.85	Flask from day 4 added to medium in 4 l spinner flask
6	4	1.5 @ 1.6×10^6	1.5	Added medium to 4 l spinner flask from day 5
7	12	3 @ 1.6×10^6	3	Flask from day 6 added to medium in 12 l New Brunswick Fermentor
8	12	6 @ 1.8×10^6	6	Added medium to 12 l New Brunswick Fermentor from day 7
9	70	12 @ 2.3×10^6	15	Fermentor from day 8 added to medium in 70 l culture vessel
10	70	27 @ 2.3×10^6	45	Added medium to 70 l culture vessel from day 9
11	200	45 @ 2.5×10^6	75	Culture vessel from day 10 added to medium in 200 l culture vessel
12	200	120 @ 1.5×10^6	50	Added medium to 200 l culture vessel from day 11
13	–	–	–	Begin harvest of 70 and 200 l culture vessel

d) Cell Growth Control

Once the "seed" culture has been introduced into the 70 l culture vessel a number of environmental parameters can now be monitored and controlled. The pH, dissolved oxygen (DO), oxidation-reduction potential (ORP) and CO_2 output can all be conti-

nuously recorded by use of strip charts. The pH, temperature, agitation rate, gas flow rate and vessel pressure can be controlled to various set-point levels.
In a typical run utilizing the BW-5147 cell line the pH is usually around 7.0 following the introduction of "seed" culture into freshly poised medium. Air is introduced through the sparge lines at a continuous rate of 1–2 lpm and 4–7 lpm, respectively, for the 70 and 200 l culture vessels. The pH is maintained at 6.95 ± 0.05 by the introduction of CO_2 on demand at a rate of 0.3 lpm and 1.2 lpm for the 70 and 200 l, respectively, or by the addition on demand of 1 molar sodium carbonate. During a typical run with BW-5147, approximately 70 ml of sodium carbonate is utilized during a 24 h period to maintain the pH in the 70 l vessel and approximately 200 ml in the 200 l vessel. It has been determined that the designated flow rate of air is sufficient to maintain the DO concentration above 20% saturation. Although DO concentration has been found to be a valuable indicator for the status of cell growth, as will be discussed presently, to date no effort has been made to control this parameter. By utilizing these empirically established control parameters we have had good success in producing large quantities of cells as will be documented in a subsequent section of this review.

e) Quality Control

Once a production run is underway there are several means of monitoring the culture to assure everything is proceeding normally. Before freshly prepared medium is utilized in a culture, the osmolality is determined. The osmolality of our modified RPMI-1640 is 320 mOSm/kg of H_2O if the medium has been supplemented with 2% fetal calf serum or 330 mOSm/kg of H_2O if 10% serum is utilized. The pH of the on-going culture is also determined by an external instrument and, if necessary, the instruments associated with the vessels are re-set. Periodically, between the period of time a culture is cut with fresh medium and the cells have reached their maximum density (late logarithmic stage) a number of other factors are examined. Each day a sample of the culture is inoculated into a blood agar plate, a tube of thioglycollate broth, a tube of Sabouraud agar and examined for mycoplasma which includes a microbiological culture procedure as well as a direct staining method [24]. If the culture is infected, one normally can see it on the blood agar plate although some organisms grow better in the other forms of detection medium utilized. The culture is also examined from the standpoint of the number of cells per unit volume of fluid and their viability. In a normal run cell viability exceeds 90%. Often it is important to determine the size distribution of cells in a culture [19]. This will indicate what percentage of the total population are approaching the logarithmic stage of growth. If the population has been maintained in the logarithmic stage of growth for several days, there will be a rather narrow size distribution. Lastly, it is important to determine the amount of product available if the product is some component other than the cells. Presently, we are interested in particular plasma membrane components of cells. Many of these components are expressed in varying amounts during different phases of the growth curve [18]. At different intervals the amount of membrane component expressed on the population of cells is examined and correlated with other parameters of cell growth.
Although the parameters utilized for quality control will obviously vary among labora-

tories we have found the aforementioned approach a workable one which gives a good indication of how a given production run is proceeding.

4. Product Acquisition

It has been our experience that multiple acquisition is the most time consuming aspect of operating a large-scale cell culture facility. We initially envisioned that the facility should be designed in such a manner that every conceivable product from cell culture could be harvested. To date we have attempted to harvest numerous products but only have definitive data on harvest times and yields for a few of these. Figure 10 illustrates the product acquisition scheme presently being followed. Plasma membrane, viruses and mediators of cellular immunity are of particular interest to our group and their acquisition will be discussed in detail.

a) Cells and Sub-Cellular Organelles

The harvest process is begun by transferring a portion of the culture from the 70 and 200 l culture vessels to the harvest vessel. In order to separate cells from the liquid the mixture is continuously fed to a Sharples Laboratory super-centrifuge equipped with a No. 1-H standard clarifier rotor at a flow rate of approximately 50 l/h. The centrifuge is enclosed in a plexiglass cabinet to contain potentially hazardous aerosols. The supernatant from the centrifuge is pumped to the supernatant holding vessel where it is held at 2–4 °C. The cells are then disrupted by use of a Stansted model AO 612 cell disruption pump equipped with a model 716 disrupting valve [25, 26]. The sub-cellular organelles and soluble enzymes are separated by differential centrifugation [27] as illustrated in Fig. 10. During the course of a normal production run, cells are harvested every 24 h and subjected to the sub-cellular fractionation scheme.

b) Viruses and Mediators of Cellular Immunity

The supernatant generated from cell harvest is held as indicated and every 48 h concentrated by use of Amicon Diafiber hollow fiber cartridges in conjunction with an air-operated dual-diaphragm pump. Viruses are separated from the fluid by use of five H10P100 cartridges whose fibers will contain molecular species larger than 100,000 daltons. This unit can process 20 l of supernatant/min and approximately 400 l is usually reduced to 2 l. The viruses contained in the concentrate (macrosolute) from the latter procedure is partially purified by continuous-flow zonal centrifugation by use of a Beckman CF-32 rotor equipped with a special core to increase capacity. The material from this run is purified further by discontinuous gradient centrifugation. The filtrate which passed through the hollow fibers can be subjected to a further concentration step by use of five H10P10 hollow fiber cartridges which retain molecules larger than 10'000 daltons. Our preliminary evidence indicates that the macrosolute from this concentration step contains various mediators of cellular immunity that lymphoblastoid cell lines have been reported to produce [28].

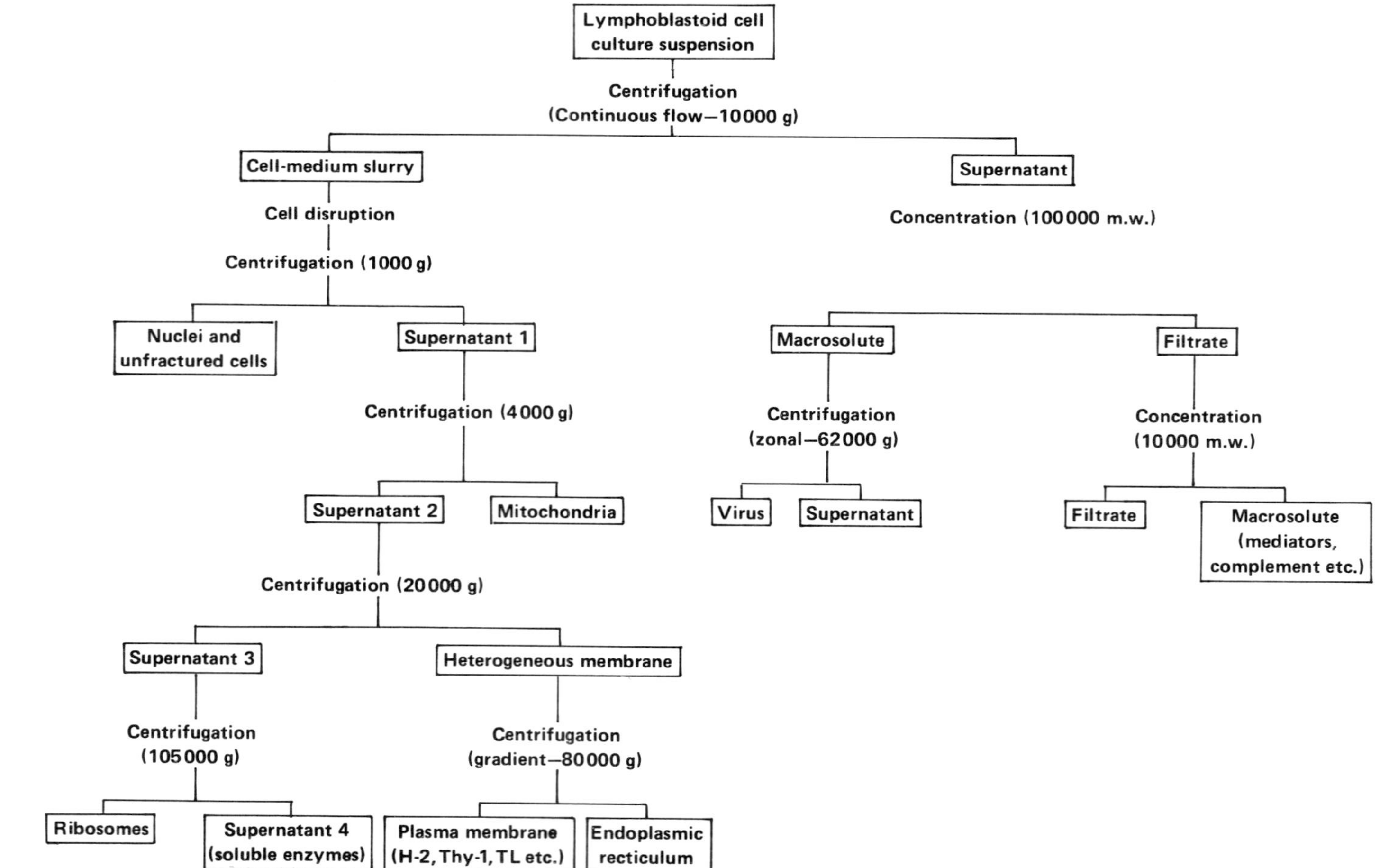

Fig. 10. Product acquisiton scheme

From these multiple steps a wealth of subcellular material and secreted cell products can be acquired for study. It is the working philosophy of our Cell Culture Center that as many components as possible from each production run be purified and made available for investigation. When looked at in this context, the cost of production is relatively inexpensive compared to the amount of material generated for study.

5. Facility Performance

a) Cell Culture Experience

It was anticipated that cell growth in larger vessels would parallel that observed in the 14 l fermentor. This expectation has in fact been realized. As can be seen in Figs. 11 and 12 the growth of BW5147 in the 70 l and 200 l culture vessels respectively, is essentially identical to that observed in the 14 l fermentor (Fig. 3). These figures depict two long-term production runs where the culture conditions were as described in Section 3 of this review. Throughout the period of culture, the pH was controlled to within a tenth of a unit. No attempt was made to control DO. A drop in the % saturation level was in general indicative of the metabolic rate of the cells. During the course of these runs, the cells often doubled every 10 h although the average doubling time was in the range of 15–20 h. There are several features of these production runs summarized in Table 3 that support our approach to large-scale growth and merits discussion. In

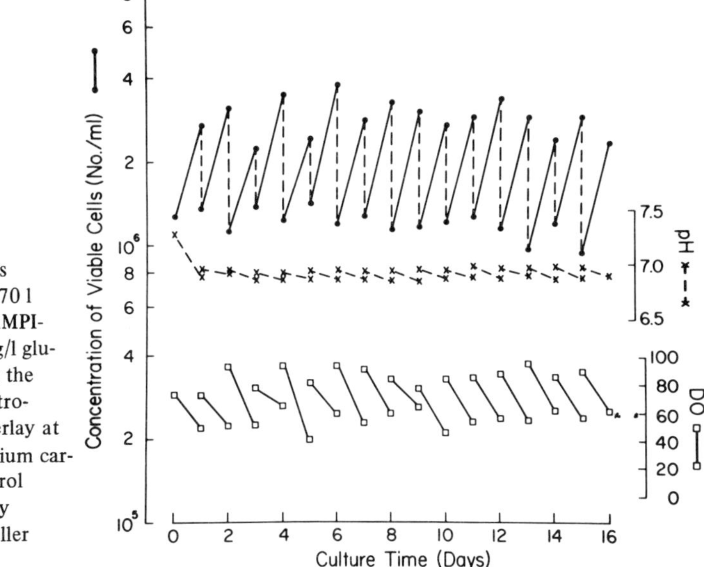

Fig. 11. Semi-continuous growth of BW5147 in a 70 l culture vessel utilizing RMPI-supplemented with 2.5 g/l glucose and 2% fetal calf as the growth medium. CO_2 introduced by sparge and overlay at 0.3 lpm and 1 molar sodium carbonate was used to control pH. Air was continuously added at 1–2 lpm. Impeller speed was 100 rpm

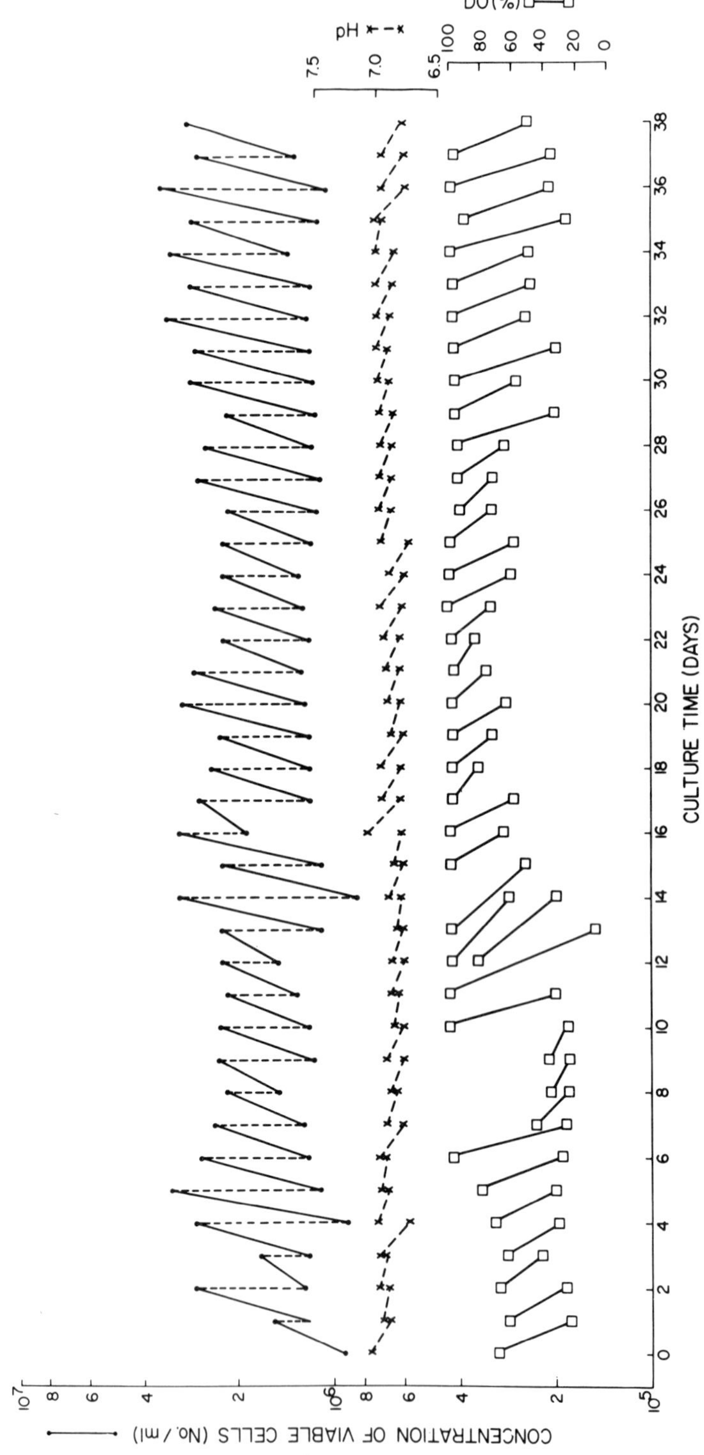

Fig. 12. Semi-continuous growth of **BW5147** in a 200 l culture vessel. Conditions were as described in Fig. 11 with the exception that CO_2 was introduced at 1.2 lpm and air at 4–7 lpm. The culture was initiated in medium supplemented with 5% horse serum and 2% calf serum. After 3 days in culture the serum level was reduced to 2% horse serum. After 29 days in culture the horse serum was replaced with 2% fetal calf in the medium for the duration of the run

Table 3. Summary of typical BW-5147 production runs

	Culture vessel	
	70 l	200 l
Length of production (days)	16	38
Medium transfers (no.)	17	42
Culture harvests (no.)	16	36
Culture harvests (l)	680	3555
Cells produced (no.)	2.04×10^{12}	9.15×10^{12}
Average density (cells/ml $\times 10^6$)	2.97	2.61
Range of densities (cells/ml $\times 10^6$)	2.24–3.82	1.53–3.56
Average cell viability (%)	95	97
Range of cell viability (%)	91–99	91–99
Operation with 2% HS (days)	–	23
Operation with 2% FCS (days)	15	10
Operation without antibiotics (days)	0	5

HS = horse serum
FCS = fetal calf serum

the 70 l culture vessel we have run for 16 consecutive days without encountering significant problems. During this period, the viability of the population was never less than 91%. The maximum density attained was 3.82×10^6 cells/ml with an average density of 2.97×10^6 cells/ml. A total of 680 l of culture medium was utilized which yielded 2.4×10^{12} viable cells. It is important to note that the cells were grown in medium supplemented with 10% fetal calf serum until they were inoculated into the 70 l vessel. From this point onward medium with only 2% fetal calf serum was utilized. We have in fact observed that many of our murine lymphoblastoid cell lines grow as well in medium plus 2% serum as in medium supplemented with larger percentages [29]. Essentially the same production rate has been attained in the 200 l culture vessel. However, as indicated this vessel has been operated for 38 consecutive days during which the last 5 days of culture was without antibiotics. We feel that this is the ultimate test of equipment design and technique of operation. Moreover, during this run we were able to grow cells in medium supplemented with 10% horse serum followed by 2% and finally with 2% fetal calf without any observed effect on viability or doubling times. During the course of this run, approximately 3800 l of medium was prepared and 3555 of this utilized to produce 9.15×10^{12} viable cells. The average daily harvest from the 200 l vessel was approximately 2.4×10^{11} cells or about twice the quantity harvested from the 70 l vessel. The cells in the 70 l vessel more consistently reached higher densities than the 200 l vessel. We feel this is not related to the size of the vessel per se but to other factors such as the degree of attention given to a given production run as opposed to another. Although we have been excited about the success of the operations to date a detailed evaluation of the data indicates that optimal growth of the BW5147 culture has not been attained. As was indicated in Section 1, a most important consideration of the semi-continuous approach to cell culture was to never let cells reach the stationary phase of growth. When this occurs, cells usually will enter a lag phase after being cut to a lower density by the addition of fresh medium. Subsequently the number of cells generated in a given period will be less than had the cells been cut at a point in time so as to remain in a logarithmic stage of growth. In any large-scale operation time and labor are most important considerations and for optimal efficiency one would like to approach a situation in which the maximum number of cells are generated with the least amount of medium in the shortest possible time. As one can readily gather from Figs. 11 and 12 the maximum density of cells attained fluxuates from day to day as does the doubling time. If the culture is monitored at closer intervals than indicated in these figures, the reason for this fluxuation becomes apparent. Figures 13–15 illustrate the data from day 6, 7, and 8 of a run in the 70 l vessel. As can be seen in Fig. 13 the cells doubled within 11 h from the time they were cut with fresh medium. However, from 12 h onward the rate of metabolism decreased as was evident by the decrease in consumption of oxygen. The cells had almost reached their maximum density of 3.8×10^6 cell/ml by 16 h. From 16 h onward the rate of oxygen consumed by the cells decreased and the CO_2 output and lactic acid production was obviously less as evidenced by the rise in pH from 6.9 to 6.98. Moreover, as can be seen in Fig. 14, the cells failed to divide further after the addition of fresh medium until 12 h later.

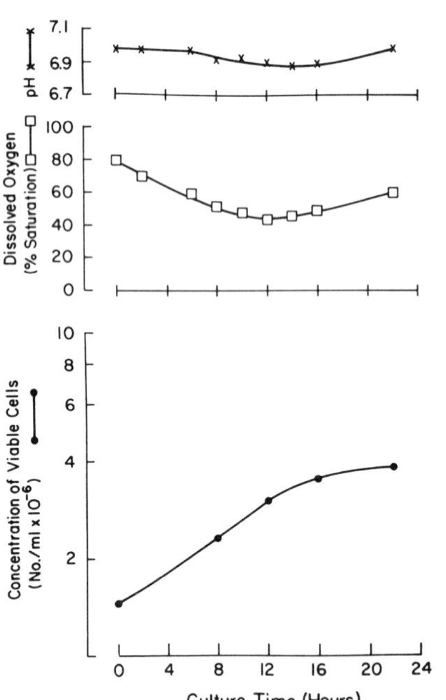

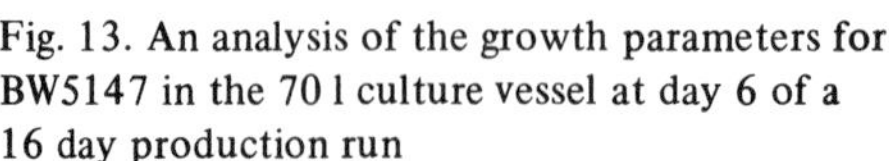

Fig. 13. An analysis of the growth parameters for BW5147 in the 70 l culture vessel at day 6 of a 16 day production run

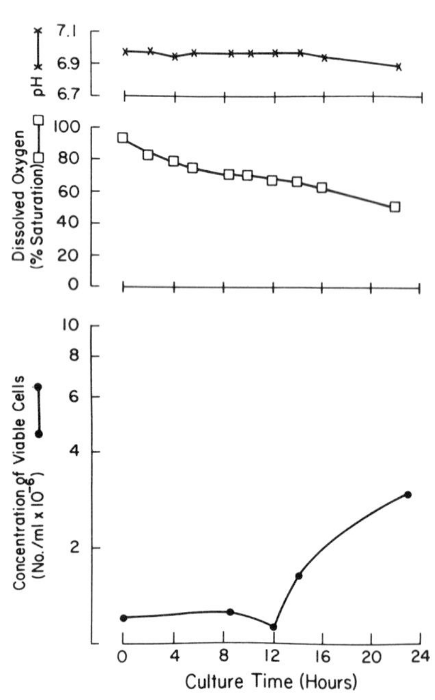

Fig. 14. An analysis of the growth parameters for BW5147 in the 70 l culture vessel at day 7 of a 16 day production run

Due to the fact that the cells reached their maximum density in the middle of the night when no one was available to conduct the harvest and add fresh medium, they entered the stationary phase of growth. The number of cells harvested on day 7 was less due to the lag time experienced as well as the fact that a density of only 2.9×10^6 cells/ml was achieved. The cells had probably not reached their maximum density when harvested as evidenced by the fact that the DO concentration and pH was still on the decline (Fig. 14). By day 8 (Fig. 15) the cells were again dividing at approximately the same rate as on day 6 and a maximum density of 3.5×10^{16} cells/ml was reached. The cells were harvested at the optimal time as judged by the slope of the DO and pH concentration curves and from the observation that on day 9 they were doubling about every 15 h.

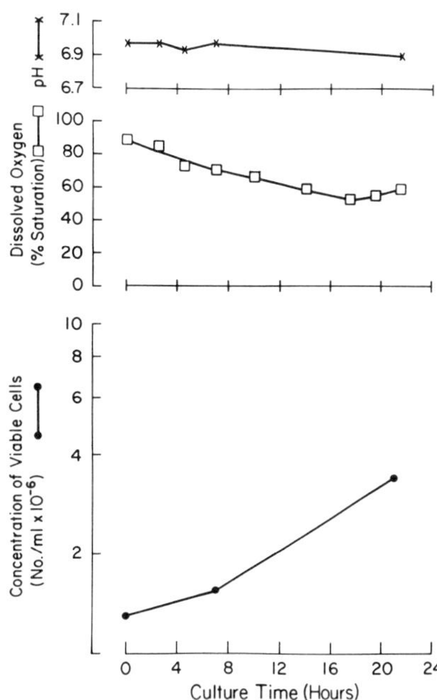

Fig. 15. An analysis of the growth parameters of BW5147 in the 70 l culture vessel at day 8 of a 16 day production run.

These data stress the need for frequent monitoring of the process and the advisability of harvesting and adding fresh medium whenever the status of the cells dictate. Although many of the problems of large-scale cell culture have been overcome in this facility it is evident from the situation just described that more precise control over environmental parameters in the culture will yield even larger numbers of cells per unit volume of medium.

b) Maintenance

Before a production run is initiated there are a number of maintenance procedures that should be conducted. It is extremely important to check and calibrate the DO and pH probes. The DO probe is only reliable for approximately 30 sterilization cycles. After this time, the sensitivity decreases and there are fluxuations in the output. The inlet and outlet air filters are also examined for leaks. We never sterilize the system until immediately before use and always try to keep the vessels under positive air pressure. This caution has allowed us to maintain vessels with liquid for as long as 60 days without any sign of microbial contamination.

With regard to breakdown time most of the problems have concerned the electronic aspect of the system, although problems with control valves and steam traps have also arisen. There have been numerous failures in our pH monitors and controls which are related to faulty circuits. The steam trap active elements were constructed of a low grade stainless steel which corroded. This was mainly due to the high resistivity (10 MΩ/cm) water used in the system. This problem has been eliminated by material selection. Early on, we were bothered with control valves that failed to operate. This problem was based on the fact that the air used to operate these valves had a high moisture content. The addition of a dryer to the air generation system has eliminated this problem. Although the primary facility has performed relatively well there are still a few problems which must be resolved to increase the efficiency of the system. There have been instances where the support equipment has failed. Most of these problems were quickly identified and eliminated. However, as more operation time is logged other faults of the facility which are now latent may appear. We feel the facility is still in a "debugging" phase and other modifications and experience will be necessary in order to generate $1-2 \times 10^{12}$ cells each 7-day work week reproducibly and efficiently.

6. Further Expectations

a) Improved Cell Growth

From the data presented it is evident that more refined culture conditions would increase the yield of cells. The semi-continuous or "feast-famine" approach to cell growth is unlikely to be the optimal system. We envision that a system whereby a cocktail of nutrients is continuously added to the culture once the cells have entered the logarithmic stage of growth would produce more cells per unit volume of medium. This procedure would also provide a continuous source of cells for harvest. Before this approach can be effected there are a number of factors which must be understood. First, is the cessation of growth as observed in Fig. 2 due to a depletion of nutrients or to the build-up of toxic metabolic by-products. From data not presented in this review [29] we have shown that the pH range utilized for the growth of BW5147 is optimal for their growth. Moreover, the DO never drops to a level which would be insufficient for growth. Therefore, we plan to analyze the medium during various stages of growth to determine if there are certain nutrients which are being depleted. Then, by further experimentation a cocktail will be formulated to be utilized in a continuous culture mode. At this point we should be able to determine the stage of growth where possible toxic metabolic by-products reach a level sufficient to inhibit growth. This approach should ultimately provide knowledge for the growth of BW5147 to densities approaching that possible in the intact animal. This cell line when grown as an ascites tumor within the appropriate mouse strain will reach densities of 1×10^8 cells/ml. By appropriate environmental control of culture conditions this level of growth should be attainable in our large-scale system.

b) Cell Product Enhancement

In a previous study we have demonstrated that the expression of a number of cell surface components fluxuates throughout the various states of growth [18, 19]. The reasons for this have not been determined. If the cells are always maintained in a logarithmic state of growth, the expression is relatively constant. Regardless of what product one is interested in the yield will probably be determined by culture conditions. The production of viruses and mediators of cellular immunity have certainly been shown to vary on a given murine lymphoblastoid cell line grown in small-scale as well as large-scale [28, 30]. As more defined conditions for cell growth in a continuous mode become available the enhancement of cellular products can be addressed.

Epilogue

It has been our aim to provide insight into the operation of a large-scale culture system for mammalian cells. This technology is rapidly evolving and should, in the near future, be equal to that utilized for microbial fermentation. The mammalian cell is obviously more complex and demanding. However, our ability to grow these cells is increasing daily. The products generated should greatly enhance investigations aimed at understanding the inner workings of mammalian cells. It should be possible to produce from cells in culture every product that cells normally generate in the body. The implication of this achievement staggers the imagination and provides the stimulus for our work.

Acknowledgements

These investigations were supported by grant No. GB-43575X from the Human Cell Biology Section of the National Science Foundation; grant No. IM 33A from the American Cancer Society, Public Health Service grant No. CA-15338 and CA-18609; Contract No. NCI-CB-43920-31 and Cancer Center Core Support Grant No. CA-13148 from the National Cancer Institute.
The comments of Dr. Nick Harakas and Dr. Paul Barstad as well as the expert technical assistance of Mike Cox is gratefully acknowledged.

References

1. Holmström, B., Hedén, C.-G.: Biotechnol. Bioeng. **6**, 419 (1964).
2. Hodge, H. M., Klein, F., Bandyopadhyay, A. K., Robinson, O. R., Jr., Shibley, G. P.: App. Microbiol. **27**, 224 (1974).
3. Mizrahi, A., Vosseller, G. V., Yzgi, A., Moore, G. E.: Proc. Soc. Exp. Biol. Med. **139**, 118 (1971).
4. Klein, F., Rosensteel, J. F., Hummer, R., Charmella, L. J.: App. and Environ. Microbiol. **31**, 995 (1976).
5. Ziegler, D. W., Davis, E. V., Thomas, W. J., McLimans, W. F.: App. Microbiol. **6**, 305 (1958).
6. Cooper, P. D., Wilson, J. N., Burt, A. M.: J. Gen. Microbiol. **21**, 702 (1959).
7. Telling, R. C., Elsworth, R.: Biotechnol. Bioeng. **7**, 417 (1965).
8. Telling, R. C., Radlett, P. J., Mowat, G. N.: Biotechnol. Bioeng. **9**, 257 (1967).
9. Radlett, P. J., Telling R. C., Stone, C. J., Whiteside, J. P.: App. Microbiol. **22**, 534 (1971).
10. Radlett, P. J., Telling, R. C., Whitside, J. P., Maskell, M. A.: Biotechnol. Bioeng. **14**, 437 (1972).
11. Fontanges, R., Deschaux, P., Beaudry, Y.: Biotechnol. Bioeng. **13**, 457 (1971).
12. Klein, F., Jones, W. I., Jr., Mahlandt, B. G., Lincoln, R. E.: Appl. Microbiol. **21**, 265 (1971).
13. Wiles, C. C., Smith, V. C.: Amer. Inst. Chem. Eng. Bioeng. Technol. **67**, 85 (1969).
14. Moore, G. E., Hasenysusch, P., Gerner, R. E., Burns, A. A.: Biotechnol. Bioeng. **10**, 625 (1968).
15. Girard, H. C., Okay, G., Kirilcim, Y., Hogan, R.: Bull. Off. int. Epiz. **79**, 255 (1973).
16. Telling, R. C., Radlett, P. J.: Adv. App. Micro. **13**, 91 (1970).
17. Horibata, K., Harris, A. W.: Exp. Cell Res. **60**, 61 (1970).
18. Zwerner, R. K., Runyan, C., Cox, R. M., Lynn, D., Acton, R. T.: Biotechnol. Bioeng. **17**, 629 (1975).
19. Zwerner, R. K., Acton, R. T.: J. Exp. Med. **142**, 378 (1975).
20. Ralph, P.: J. Immunol. **110**, 1470 (1973).
21. Acton, R. T., Barstad, P. A., Cox, R. M., Zwerner, R. K., Wise, K. S., Lynn, J. D.: In: Cell Culture and its Application (R. T. Acton and J. D. Lynn Eds.). New York: Academic Press 1977.
22. Lynn, J. D., Acton, R. T.: Biotechnol. Bioeng. **17**, 659 (1975).
23. Cameron, J., Godfrey, E. I. J.: Appl. Bact. **31**, 405 (1968).
24. Low, I. E.: Health Laboratory Sci. **13**, 129 (1976).
25. Wright, B. M., Edwards, A. J., Jones, V., E.: J. Immunol. Methods **4**, 281 (1974).
26. Runyan, C. C., Acton, R. T.: Fed. Proc. **34**, 1013 (1975).
27. Allan, D., Crumpton, M. J.: Biochem. J. **120**, 133 (1970).
28. Goldstine, S. N., Acton, R. T.: Unpublished observations.
29. Barstad, P. A., Acton, R. T.: Unpublished observations.
30. Wise, K. S., Acton, R. T.: Unpublished observations.

A Complementary Approach to Scale-Up

Simulation and Optimization of Microbial Processes

S. Aiba and M. Okabe
Biochemical Engineering Laboratory, Institute of Applied Microbiology,
University of Tokyo, Japan

Contents

Summary

A typical batch process may be subdivided into:
a) continuous sterilization of culture medium
b) aeration and agitation in bioreactor vessel
c) ancillary operations and equipment.

Assuming that the accumulation rate of specific product in the broth and the annual output of the product are given, objective functions defined by the sum of annual expenditures on utilities and equipment may be simulated, and optimized via the golden section method with respect to a) and b).

The modified complex method may be applied to minimize the objective function for the whole batch process, using as design variables the working volume of bioreactor, process time and boiler capacity.

Optimization becomes increasingly beneficial as the magnitude of the objective function increases. This means that optimization of each individual operation or item of equipment in the process (i. e., suboptimization which disregards the mutual relationships among items constituting the process) becomes less important from the viewpoint of process synthesis. The significance of the "synthetic" approach, complementary to the usual "analytical" approach to scale-up is emphasized.

1. Introduction

During the past three decades many workers in industrial microbiology have attempted to find rules which permit a successful scale-up of microbial processes. The volumetric coefficient of oxygen transfer in an aerobic fermentation, power input of agitation per unit volume of broth, and so forth have been presented by these investigators as useful indices of scale-up [4, 12, 16]. If vessels are designed so that these indices are equated for reactor vessels of various capacities, successful scale-up implies that the bioreaction performance observed in small vessels can be realized in larger ones.

Recently, kinetic studies on various microorganisms have been conducted by many workers, and this trend may be beneficial for the future exploration of scale-up problems still remaining to be solved [8–10, 19]. The above-mentioned studies and those related directly to scale-up indices might well be termed "analytical" in nature.

In contrast to these analytical methods this paper describes a synthetic approach to process design, in which the most economical design is sought, taking into account interactions among various items of equipment and operations of the process (see Fig. 1) for overall optimization.

A typical microbial process, shown in Fig. 1, will be considered [14]. Though substantially batchwise in operation except for continuous sterilization of the culture medium, the microbial process here excludes, for simplicity, the product separation and subsequent processes. However, the basic equipment items for steam generation, sterile air supply and so forth are included in this process scheme.

First we shall consider the simulation of a specific operation in the process such as continuous sterilization of the medium, and, second, the process as a whole will be optimized.

It is difficult to judge mathematically whether or not convergence of the objective function (sum of annual expenditures on utilities and equipment) for the whole process represents an optimum, because of complexities of multivariables in the process design–represented in Fig. 1. Consequently, the simulation of some of the principal operations mentioned above is needed to check directly whether or not a suboptimal condition exists in any operation, though this preexamination does not always lead directly to determining optimal conditions for the process as a whole.

In view of the nature of "synthesis", cost estimates for investment from item to item in Fig. 1 and for the utilities electricity, steam, water, and air are needed. In the light of the character of this paper, however, details of cost estimation which are described elsewhere will be omitted here, and only system equations relevant to the principal operations will be discussed, to show the basis of cost estimation [2, 3].

To recapitulate, the purpose of this paper is to present a synthetic approach to scale-up in biotechnology employing several independent design variables.

Although the procedures used to derive system equations for the process and to assess the objective functions will be considerably simplified from case to case (see later) and there is scope for further refinement [15] (for example, sensitivity analysis), the method has potential application in the industry, especially when viewed as complementary to the "analytical" approaches to scale-up previously published by many workers.

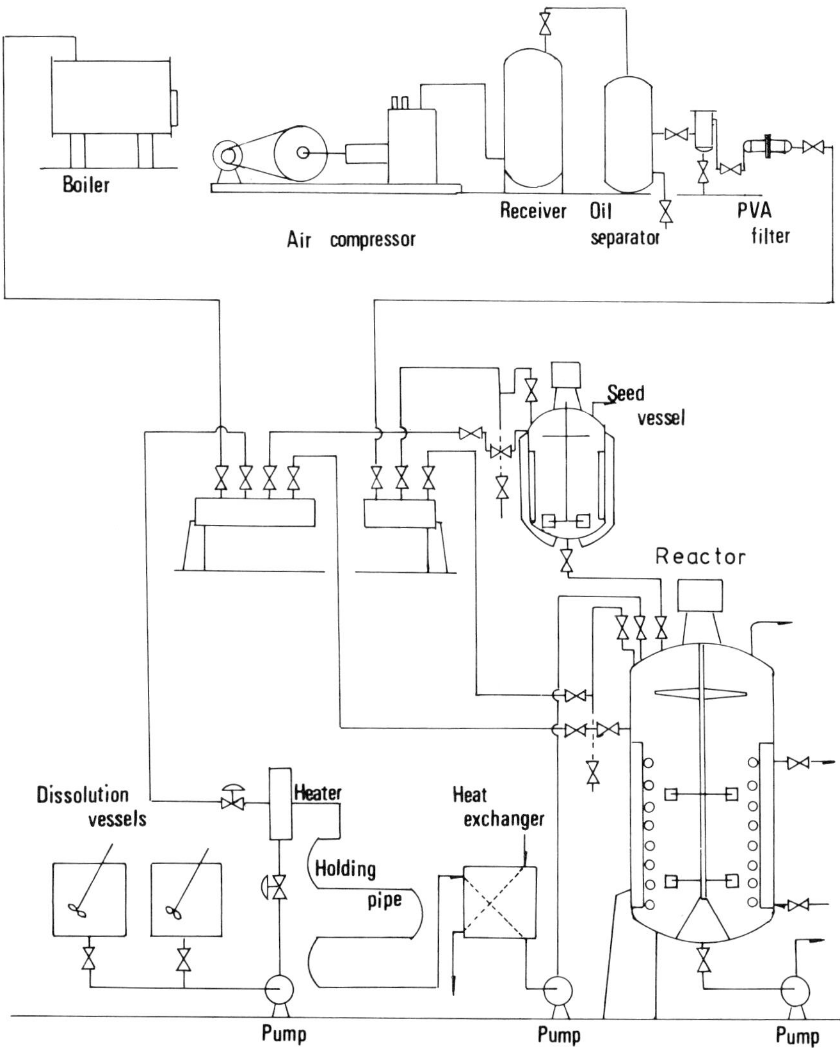

Fig. 1. Flow diagram of a microbial process

2. Process-Modelling and System Equations

2.1 Time-Cycle of Operations

The time-cycle of the various operations, from the cleaning of the reactor vessel to the discharge of the broth in Fig. 1, is needed in order to define the annual number of runs and to secure an average "rate" of operation.

The cycle time, t_C for each batch run is given by:

$$t_C = t_{01} + t_{02} + t_{03} + t_{04} + t_F + t_{05}, \tag{1}$$

where
t_F = time for reaction (h)
t_{01} = time for cleaning the reaction vessel (h)
t_{02} = time required for sterilization of the empty vessel (h)
t_{03} = time for continuous sterilization of the culture medium (h)
t_{04} = time needed for inoculation (h)
t_{05} = time for discharge of broth (h).
Designating P_A, t_A, V, N_F, and C_P as the annual amount of specific product in broth, annual working period, volume of broth in the vessel, number of vessels, and product concentration in the broth, respectively, the annual number of runs, N_C is:

$$N_C = \frac{P_A}{VC_pN_F} = \frac{t_A}{t_C}. \tag{2}$$

For convenience, the number of vessels will be taken hereafter as one ($N_F = 1$). This simplification does not infringe upon the generality of discussion, because more vessels may be employed when the value of V suggested from the optimization appearing later on is too large for the construction of a single vessel.
From Eqs. (1) and (2),

$$t_{03} = (t_A/N_C) - (t_{01} + t_{02} + t_{04} + t_F + t_{05}). \tag{3}$$

If the value of P_A is predetermined and, also, if C_P as a function of t_F is given a priori, Eq. (2) indicates that N_C depends on both V and t_F.
Since the values of t_{01}, t_{04}, and t_{05} can all be taken as 1.0 h, and t_A as 7200 h, for simplicity, Eq. (3) suggests that t_{03} is a function of V, t_F and S_B, taking for granted the fact that t_{02} required for blank sterilization by blowing live steam is another function of V and S_B (see below), where S_B is boiler capacity (ton h^{-1}).
The heat balance for blank sterilization of the fermentation vessel is:

$$W\frac{dT}{dt} = \alpha Q_s \cdot \{\lambda + (T_s - T)c\} - A_F U_F (T - T_o), \tag{4}$$

where:
A_F = outer surface area of vessel (m^2)
c = specific heat of water (kcal ton^{-1} °C^{-1})
Q_s = rate of steam supply (ton h^{-1})
T = temperature of vessel (°C)
T_o = room temperature (°C)
T_s = steam temperature (°C)
U_F = overall heat-transfer coefficient for vessel wall due to natural convection, (kcal $m^{-2}h^{-1}$ °C^{-1})
W = heat capacity of vessel (kcal °C^{-1})

α = fraction of steam condensed
λ = latent heat of vaporization of water (kcal ton^{-1}).
Clearly, W in the equation depends on the total volume of the vessel, which was taken as $V/0.70$. Then, the time, t_{021}, required for raising the temperature of the vessel from $T = T_o$ to a specific value of T for sterilization and another time, t_{023}, for the vessel to be cooled down after cessation of steam supply can be estimated from Eq. (4), once the values of α for the former steam-blowing and working volume, V, of the reactor are assumed.
Thus

$$t_{02} = t_{021} + t_{022} + t_{023}, \tag{5}$$

where:
t_{022} = period of time during which temperature of reactor is kept at $T = T$; t_{022} was taken as 0.5 h.
Now, it is clear that the values of both S_B and V affect the value of t_{02} in Eqs. (5) and (3); independent variables to define the time, t_{03}, for continuous sterilization of culture medium are V, t_F, and S_B.

2.2 Continuous Sterilization of Culture Medium

2.2.1 Holding Pipe

The degree of sterilization, $(n/n_o)_{x=L}$, at the exit of the holding pipe is shown in steady state by the following equation [1].

$$(n/n_o)_{x=L} = \frac{4\zeta \exp\left(\frac{PeB}{2}\right)}{(1+\zeta)^2 \exp\left(\frac{PeB}{2}\zeta\right) - (1-\zeta)^2 \exp\left(-\frac{PeB}{2}\zeta\right)}, \tag{6}$$

where:
n = number concentration of contaminants in fresh medium after sterilization (ml^{-1})
n_o = number concentration of contaminants in raw medium before sterilization (ml^{-1})
L = length of holding pipe (m)
PeB = Bodenstein number $(= \bar{u}L/E_z)$
E_z = axial dispersion coefficient of culture medium in holding pipe (m^2 s^{-1})
$\bar{u}$ = mean linear velocity of culture medium through holding pipe (m s^{-1})
x = distance in axial direction
$\zeta = \left(1 + 4\frac{Nr}{PeB}\right)^{1/2}$
$Nr = kL/\bar{u} = k \cdot t_\theta$
k = death-rate constant of contaminants (s^{-1})
t_θ = mean residence time of culture medium in holding pipe (s).
The death-rate constant, k, is given by the Arrhenius equation as follows [1]:

$$k = \alpha' \exp\{-E/R(T+273)\} = \alpha' \exp\{-\beta'/(T+273)\}, \tag{7}$$

where:
E = activation energy (cal mol^{-1})
R = gas constant, 1.98 cal mol^{-1} K^{-1}
T = sterilization temperature (°C)
α' = empirical constant
β' = empirical constant $(= E/R)$.
From the definition of PeB,

$$L = PeBd\left(\frac{E_z}{\bar{u}d}\right), \tag{8}$$

where
d = diamater of holding pipe (m).
Assuming that cooling is effected simultaneously with the completion of passage of medium through the holding pipe, independent variables to design the continuous sterilizer can be taken as the Bodenstein number and sterilization temperature, provided the degree of sterilization is given.
If the value of t_{03} is assessed from Eqs. (2) to (4) for given values of V, t_F, and S_B, the diameter, d, and length, L, of the holding pipe can be designed in the following steps: With given values of PeB and T,
i) Assume a value of d and determine $\bar{u}$ $(= V(\pi/4)d^2 t_{03})$
ii) Calculate the Reynolds number, $N_{Re}(= d\bar{u}\rho/\mu$, where ρ and μ are the density and viscosity of culture medium at T)
iii) Estimate $E_z/\bar{u}d$ from the experimental correlation between $E_z/\bar{u}d$ and N_{Re} [11]
iv) Calculate L from Eq. (8)
v) Estimate the value of Nr $(= kL/\bar{u})$ from Eq. (6) under the given conditions of $(n/n_o)_{x=L}$ and PeB [1]
vi) Calculate k from Eq. (7), followed by another calculation of L from Nr and $\bar{u}$ secured in v) and i), respectively
vii) Repeat steps i) to vi) changing d values until both values of L obtained in iv) and vi) agree within an allowable tolerance.

If no heat loss is assumed from the outer surface of the holding pipe, the rate of steam injection into the pipe, Q_s ton h^{-1}, is:

$$Q_s = \frac{c\rho \dfrac{V}{t_{03}}(T - T_o)}{\lambda + (T_s - T)c}, \tag{9}$$

where
c = specific heat of culture medium; assumed equal to that of water (kcal ton^{-1} °C^{-1})
T_o = room temperature (°C)

2.2.2 Heat Exchanger

The heat-transfer area of the heat exchanger is given by:

$$A = \frac{c\rho \dfrac{V}{t_{03}}(T - T_F)\ln\left(\dfrac{T - T_x}{T_F - T_c}\right)}{U\{(T - T_x) - (T_F - T_c)\}}, \tag{10}$$

where
A = heat-transfer area of heat exchanger (m^2)
T_c = temperature of cooling water at the inlet of heat exchanger (°C)
T_F = culture-medium temperature at exit of heat exchanger (°C)
T_x = temperature of cooling water at exit of heat exchanger (°C)
U = overall heat-transfer coefficient (kcal m^{-1} h^{-1} °C^{-1}).
Besides V, t_F, and S_B which affect the value of t_{03} as mentioned earlier, T and T_x will be taken as independent variables in Eq. (10) when values of U, T_F, and T_c in the equation are fixed.
The rate of cooling-water supply, Q_w ton h^{-1} through the heat exchanger is given by:

$$Q_w = \frac{c\rho \frac{V}{t_{03}} (T - T_F)}{(T_x - T_c)c}. \tag{11}$$

2.3 Aeration and Agitation in the Reactor Vessel

Following the suggestion in papers published by earlier workers that the volumetric oxygen transfer coefficient, K_v, multiplied by the partial pressure of oxygen, p, in the reactor is most probably a significant factor affecting the concentration of fermentation product in a broth [1, 4], it is assumed in the following discussion that C_p can be expressed as a polynomial function of t_F when $K_v p$ as parameter is kept at a specific value throughout.
Many workers have presented empirical formulae for K_v as a function of operating variables, such as aeration rate and speed of impeller. In addition, K_v values are subject to change depending on temperature, existence of organic substance and surface-active agents in the broth, etc. [1]. However, the value of K_v under given conditions may be assumed to depend only on the aeration rate and agitation speed, and the formula published originally by Cooper et al. [5] will be employed, i. e.:

$$K_v = 0.0635\,(P_g/V)^{0.95}\,v_s^{0.67}, \tag{12}$$

where
P_g = power consumption of agitator under aerated conditions (HP)
v_s = superficial air velocity based on empty cross-sectional area of vessel (m h^{-1}).
Assuming that the pressure inside the vessel is kept at 0.2 kg cm^{-2} gauge, the average pressure of oxygen, p, is given by:

$$p = \frac{1.2 + \left(1.2 + \frac{D_T}{10.3}\right)}{2} \cdot 0.21, \tag{13}$$

where
D_T = diameter of vessel (m).
If the liquid depth in the reactor is twice the vessel diameter,

$$D_T = \left(\frac{4}{\pi} \cdot \frac{1}{2} V\right)^{1/3} = 0.86\, V^{1/3}. \tag{13'}$$

From Eqs. (13) and (13′),

$$p = 0.252 + 0.018\ V^{1/3}. \tag{14}$$

But

$$v_s = \frac{Q_A \cdot 60}{(\pi/4) \cdot D_T^2} = \frac{Q_A \cdot 60}{(\pi/4)\,(0.86\ V^{1/3})^2} = (1.03 \times 10^2)\frac{Q_A}{V^{2/3}}, \tag{15}$$

where
Q_A = air flow rate ($m^3\ min^{-1}$).
Cancelling out v_s from Eqs. (12) and (15), followed by derivation of the term $K_v p$ from Eqs. (12) and (14), P_g is given by the following equation, provided $K_v p$ is kept constant at the specified value throughout:

$$P_g = \left\{\frac{\text{const.}}{Q_A^{0.67}\,(0.358\ V^{-1.4} + 0.026\ V^{-1.1})}\right\}^{1.05}. \tag{16}$$

It can be seen from Eq. (16) that P_g is dependent on Q_A, once the value of V is given. Here, it is necessary to point out that the equation simply shows an increase in P_g when the value of Q_A decreases, and vice versa, without elaborating on the contribution of the impeller to the value of P_g.
An empirical formula presented by Fukuda et al. [7] for the power consumption of agitators under aeration is cited below, though the formula is not dimensionally sound.

$$P_g = (2.4 \times 10^{-3}) \left(\frac{P_o^2 N' D_i'^3}{Q_A'^{0.08}}\right)^{0.39}, \tag{17}$$

where
D_i' = impeller diameter (cm)
N' = rotation speed of impeller (rpm)
P_o = power consumption of agitator without aeration (HP)
Q_A = air flow rate ($cm^3\ min^{-1}$).
The units of P_o, N', D_i', and Q_A' in the above equation are first converted, respectively, to P_o' (kg m s^{-1}), N (rps), D_i (m), and Q_A ($m^3\ min^{-1}$). Second, the ratio of D_i to D_T is taken as 1/3 and the Power number is assumed to be 6 [1]. Then,

$$D_i = (1/3)\,D_T = (1/3)\,(0.86)\,V^{1/3} = 0.29\ V^{1/3} \quad \text{[cf. Eq. (13′)]}, \tag{18}$$

$$\frac{P_o' g_c}{N^3 D_i^5 \rho} = 6, \tag{19}$$

where
g_c = conversion factor, kg m $kg^{-1}\ s^{-2}$.
The conversion of each term in Eq. (17) to consistent units gives:

$$P_g = (5.7 \times 10^{-2})\,P_o'^{0.78} N^{0.39} D_i^{1.17} Q_A^{-0.031}. \tag{20}$$

Cancelling out P_o' from Eqs. (19) and (20), and also using Eq. (18) and $\rho = 10^3\,\mathrm{kg\,m^{-3}}$,

$$P_g = (1.6 \times 10^{-2})\, N^{2.73} V^{1.69} Q_A^{-0.031}. \tag{21}$$

At first glance Eqs. (16) and (21) seem inconsistent, because the effect of Q_A on P_g in the former equation is far more marked than in the latter. However, this apparent inconsistency originates because $K_V p$ is taken as constant in Eq. (16), whilst Eq. (21) is derived independently of Eq. (12), which is the basis of Eq. (16).
It is seen from Eq. (21) that the effect of N on P_g is much larger than that of Q_A.

3. Procedure of Simulation and Optimization

The process in Fig. 1, comprising various operations and items of equipment, is subdivided for ease of simulation and optimization into the following groups:
i) continuous sterilization of the medium
ii) aeration and agitation in the vessel
iii) ancillary operations and equipment, other than i) and ii).
The objecitive function, ϕ is defined by:

$$\phi = M_c N_C + \sum_i e_i E_i N_C + \alpha_L \sum_m I_m \tag{22}$$

$$\left\{\left(M_c + \sum_i e_i E_i \;\; N_C + \alpha_L \sum_m I_m\right\}_1 + \sum_{i'=2}^{3} \left\{\left(\sum_i \; e_i E_i\right) N_C + \alpha_L \sum_m I_m\right\}_{i'}$$

$$= \phi_1 + \phi_2 + \phi_3, \tag{23}$$

where
E_i = i-th utility required per batch (i: electricity, steam, water, and air)
e_i = cost of i-th utility
I_m = investment cost of m-th equipment (m: holding pipe, heat exchanger, reaction vessel, boiler, etc.)
M_c = cost of fresh medium per batch
α_L = depreciation (taken as 0.3) multiplied by Lang factor (= 3.6) [6]
ϕ_1 = objective function for continuous sterilization of culture medium (cost/year)
ϕ_2 = objective function for aeration and agitation in fermentation vessel (cost/year)
ϕ_3 = objective function for ancillary operations and equipment (cost/year).
Here again for simplicity, the costs of pumps, valves, piping, etc. are excluded.
The fact that the annual cost of raw medium is included in the first subgroup has no particular significance. Referring to the first operation (Subsystem 1) I_m is for the holding pipe and heat exchanger. When the values of V, t_F, and S_B are assumed, the discussion in previous sections permits, from the system Eqs. (3) and (6) to (8), assessment of d and L for the holding pipe at a predetermined degree of sterility, taking PeB and T as independent variables. Likewise, the basis for estimating I_m for the heat exchanger is given by Eq. (10), taking T_x and T also as independent variables.

Utilities of Subsystem 1 are given by Eq. (9) for steam consumption (Q_s) and by Eq. (11) for cooling water (Q_w); values of Q_s and Q_w are affected by T and T_x, respectively, and t_{03} value in these equations by V, t_F, and S_B.
Then

$$\phi_1 = \phi_1(V, t_F, S_B, PeB, T, T_x). \tag{24}$$

As far as aeration and agitation in the bioreactor (Subsystem 2) is concerned, I_m pertains to vessel construction and agitator installation; the basis of cost estimation for the former is the value of V per se, whilst that for the latter is P_g, the value of which is given by either Eq. (17) or Eq. (21). In addition, I_m in this subsystem is assumed to incorporate the costs of the air compressor and air filter. The cost estimation for both items is based on Q_A [14]. Utilities are electricity, air, and steam. Electricity and air expended can be estimated from P_g, Q_A, and t_F, and steam from Eq. (4). Accordingly,

$$\phi_2 = \phi_2(V, t_F, S_B, Q_A). \tag{25}$$

Lastly, I_m for Subsystem 3 covers the dissolution vessels, preculture vessel (see Fig. 1) and boiler. If the working volumes of these vessels are taken as proportional to V (e. g., working volume of dissolution vessel = $V/2$ and that of seed vessel = $V/100$), I_m values in this subsystem are estimated from V and S_B. For utilities, electricity requirements for agitation in the dissolution vessels are estimated by assuming that power input (= 0.5 HP m^{-3}) continues for the same period as t_{03} (dependent on V, t_F, and S_B). Utilities for seed precultivation are disregarded. It is evident from the above argument that

$$\phi_3 = \phi_3(V, t_F, S_B). \tag{26}$$

From Eqs. (23) to (26)

$$\begin{aligned}\phi = {} & \phi_1(V, t_F, S_B, PeB, T, T_x) \\ & + \phi_2(V, t_F, S_B, Q_A) + \phi_3(V, t_F, S_B).\end{aligned} \tag{27}$$

It is seen that ϕ is dependent on seven variables and the optimization requires, under proper search ranges and constraints (see later) a specific assortment of values of these variables to minimize ϕ.
It is noted from Eq. (27) that the variables V, t_F, and S_B are shared in each subsystem. Herein lies the origin of the breakdown procedure, in which each subsystem is optimized first for given values of V, t_F, and S_B, and, second, another search for optimal values of these variables is conducted on the second level, in which the separately optimized information for each subsystem accrued from the first level, for PeB^*, T^*, T_x^*, and Q_A^* is fed back so that ϕ [= $\phi_1(V, t_F, S_B, PeB^*, T^*, T_x^*) + \phi_2(V, t_F, S_B, Q_A^*) + \phi_3(V, t_F, S_B)$] is minimized.

3.1 Simulation

For convenience of computation, ϕ_1 is subdivided again as follows:

$$\phi_1 = \phi_{11}(V, t_F, S_B, PeB, T) + \phi_{12}(V, t_F, S_B, T, T_x), \tag{28}$$

where

ϕ_{11} = objective function for holding pipe (cost/year)

ϕ_{12} = objective function for heat-exchanger (cost/year).

For a given set of values of V, t_F, and S_B, simulation (and optimization) of ϕ_1 turns out to be a problem of two variables. However, a one-variable search (the golden section method [17]) is employed consecutively here for simplicity [14], taking temperature of sterilization as parameter. The golden section method is also applied to simulation (and optimization) of ϕ_2 [see Eq. (25)].

3.2 Optimization

It is clear that a multivariable search (with three coordination variables, V, t_F, and S_B) is required on the second level in optimizing the value of ϕ. The modified complex method [13] was coupled in this search with previous separate optimizations of ϕ_1 and ϕ_2 (see Fig. 2).

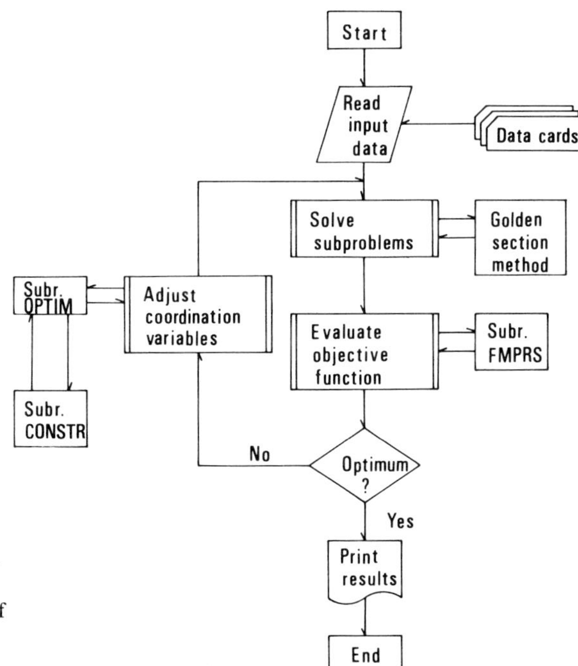

Fig. 2. Flow diagram of simulation and optimization, where CONSTR and FMPRS indicate subroutines of constraints and optimization of each operation, respectively

Explicit constraints for each independent variable in this optimization, amounting to seven in total, are needed first. These constraints will be considered in the example (see Table 1). Secondly, inequality and implicit constraints are as follows:

i) $$0 < t_{03} \leqslant (t_{03})_{max}. \tag{29}$$

The above constraint implies that the time needed for continuous sterilization shall not exceed a specific value to avoid malfunctioning of the whole process, and the value of $(t_{03})_{max}$ actually refers to practice.

ii) $$A_F U_F (T - T_o) < \alpha Q_s \{\lambda + (T_s - T) c\}. \tag{30}$$

This constraint is self-explanatory from Eq. (4) to keep the temperature of the bioreactor at T in the blank sterilization.

iii) $$c\rho \frac{V}{t_{03}} (T - T_o) < 0.8\, S_B \{\lambda + (T_s - T) c\}. \tag{31}$$

Referring to Eq. (9), the above constraint is for the temperature of culture medium to be raised instantly from T_o to T at the inlet of the holding pipe; 0.8 on the right-hand side implies the loss of steam during transportation to the extent of 20%.

4. Example of Simulation and Optimization

Supposing that a cumulative pattern of specific product in the broth at $K_v p = 0.11$ kg mol 0_2 m^{-3} h^{-1} is given [18] as shown below, the simulation and optimization of the continuous sterilization and aeration in the culture, followed by process optimization, will be exemplified (HITAC 8700/8800, Computer Center, Univ. of Tokyo).

$$\left.\begin{array}{c} C_P = 8.592 - 0.232\, t_F + 0.00329\, t_F^2 - 0.0000125\, t_F^3 \\ 72 \leqslant t_F \leqslant 132 \end{array}\right\}. \tag{32}$$

A schematic flow sheet employed is shown in Figure 2. The simulation of each operation and stagewise optimization is carried out in the subroutines (first level). The computation is iterated until the stopping criterion defined by (1.0 – ratio of present to preceding values of each objective function) become smaller than the convergence tolerance, ϵ. Values of ϵ adopted in each search will be shown later.

The modified complex method is applied in Subroutine "OPTIM" on the second level, checking whether the explicit and implicit constraints are satisfied. This search is stopped when the two stopping criteria, one defined by the relative difference of the largest and smallest values of ϕ at the vertices of the complex, and the other by the ratio of the average value of the distance between adjacent vertices to the norm of the centroid, are smaller than convergence tolerances, ϵ_1 and ϵ_2 [13].

Design data and constraints used in this example are summarized in Table 1. Cost data for utilities and empirical equations to estimate investment cost for equipment are not

Table 1. Design data and constraints used in simulation and optimization

Design data

i) Overall

annual amount required of specific product in broth, $P_A = 5 \times 10^4$ kg
annual working time, $t_A = 7.2 \times 10^3$ h

ii) Blank sterilization of bioreactor

specific heat of water, $c = 10^3$ kcal ton^{-1} °C^{-1}
temperature of vessel in blank sterilization, $T = 120$ °C
room temperature, $T_o = 20$ °C
steam temperature, $T_s = 164$ °C (6 kg cm^{-2} gauge)
overall heat-transfer coefficient for vessel wall due to natural convection, $U_F = 8$ kcal m^{-2} h^{-1} °C^{-1}
fraction of steam condensed in blank sterilization, $\alpha = 0.05$
latent heat of vaporization of water, $\lambda = 494 \times 10^3$ kcal ton^{-1}

iii) Continuous sterilization of culture medium

specific heat of culture medium or water, $c = 10^3$ kcal ton^{-1} °C^{-1}
number concentration of contaminants in fresh medium at outlet of holding pipe, $n = 10$ ml^{-1}
number concentration of contaminants in fresh medium before sterilization, $n_o = 4 \times 10^8$ ml^{-1}
temperature of cooling water at inlet of heat exchanger, $T_c = 20$ °C
culture-medium temperature at exit of heat exchanger, $T_F = 24$ °C
room temperature, $T_o = 20$ °C
steam temperature, $T_s = 164$ °C
overall heat transfer coefficient for heat exchanger, $U = 1.5 \times 10^3$ kcal m^{-2} h^{-1} °C^{-1}
empirical constant, $\alpha' = 3.11 \times 10^{62}$ min^{-1} [20]
empirical constant, $\beta' = 5.71 \times 10^4$ K
latent heat of vaporization of water, $\lambda = 494 \times 10^3$ kcal ton^{-1}
viscosity of culture medium at sterilization temperature, $\mu = 4 \times 10^{-4}$ kg m^{-1} s^{-1}
density of culture medium, $\rho = 1$ ton m^{-3}

Explicit constraints

i) Second level

boiler capacity, S_B ton h^{-1}
$7 \leqslant S_B \leqslant 20$
actual reaction time, t_F h
$80 \leqslant t_F \leqslant 130$
nominal volume of bioreactor, $V/0.7$ m^3
$100 \leqslant V/0.7 \leqslant 300$

ii) First level

inner diameter of holding pipe in continuous sterilizer, d m
$0.05 \leqslant d \leqslant 1.0$
Bodenstein number, PeB
$20 \leqslant PeB \leqslant 100$
temperature of continuous sterilization, T °C
$115 \leqslant T \leqslant 130$
temperature of cooling water at exit of heat exchanger, T_x °C
$25 \leqslant T_x \leqslant 60$
air flow rate, Q_A m^3 min^{-1}
$0.04 \leqslant Q_A/V \leqslant 1.0$

Inequality and implicit constraints

See Formulae (29) to (31) in the text, provided $(t_{03})_{max}$ in (29) was taken as 12 h included in the table, though computations of ϕ, ϕ_1 and so forth via the cost estimation are conducted elsewhere [2, 3]. In this context, values of ϕ, ϕ_1 and so on which appear next will be shown in relative scale.

5. Result and Discussion

The value of ϕ_{11} is plotted against PeB in Figure 3, the parameter being the sterilization temperature. Although the data points in the figure are calculated rather than experimental, symbols are used for clarity.

Because of the computational stipulation that values of P_A and t_A are pre-fixed and those of V, t_F, and S_B are assumed, the value of t_{03} is also given from Eqs. (2) and (3). In addition, remembering that the degree of sterility is given a priori, the solution of Eq. (6) shows that decrease of PeB is accompanied by increase of Nr ($= kL/\bar{u}$) and vice versa [1]; at a specific temperature, T, for which the k value is given by Eq. (7), the decrease of PeB entails the increase of $L/\bar{u}$.

The value of E_z will be assumed approximately constant [11] for the following discussion. Since $PeB = (\bar{u}\, d/E_z)\,(L/d) = (4/\pi)\,(V/t_{03})\,(1/d)\,(1/E_z)\,(L/d) = \text{const} \cdot (L/d^2)$,

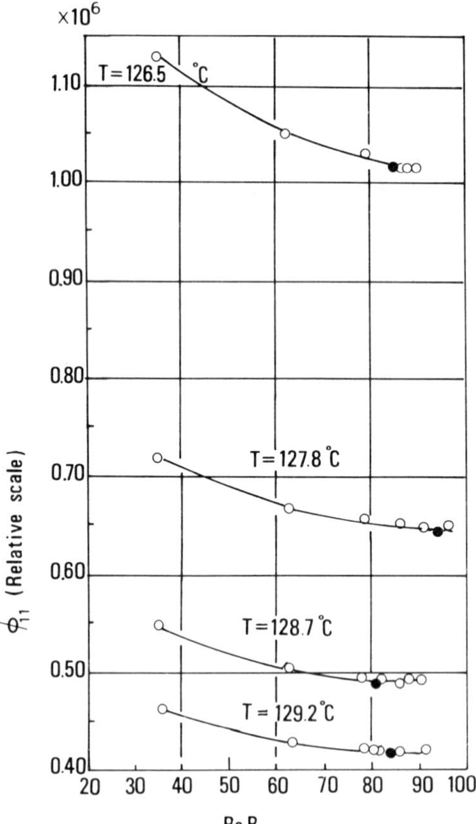

Fig. 3. Simulation of holding pipe in continuous sterilization, provided: $V = 140\ m^3$, $t_F = 116$ h, and $S_B = 8$ ton h^{-1}

decrease of *PeB* corresponds to that of L/d^2. On the other hand, the decrease of *PeB*, which entails increase of $L/\bar{u}$ as referred to above, means the increase of $L d^2$ and vice versa. Thus, the following directions among values of *PeB*, L/d^2, and $L\,d^2$ are expected.

PeB	L/d^2	$L d^2$
decrease	decrease	increase
increase	increase	decrease.

The above scheme suggests that decrease of *PeB* implies a holding pipe of larger diameter, whereas increase of *PeB* implies a pipe of smaller diameter.

So far as each curve of Figure 3 at constant value of T is concerned, the above discussion suggests that ϕ_{11} tends to increase with decrease of *PeB* primarily because of the increased investment cost for pipe, whilst the fact that ϕ_{11} values level off when *PeB* values exceed 80 in the figure might originate from the decrease in pipe diameter.

If *PeB* is fixed, the value of L/d^2 is also fixed, as is apparent from the earlier discussion; when T increases, $L/\bar{u}$ decreases, because $k\,L/\bar{u}$ = const. at constant value of *PeB*. Therefore, the decrease of $L/\bar{u}$, i. e., of $L\,d^2$, implies the decrease of both L and d. Accordingly, the observation in the figure that ϕ_{11} values decrease conspicuously when T increases at constant value of *PeB* is plausible. Savings of investment cost for pipe by the increase of T overwhelm the steam-cost increase as expected from Eq. (9). The solid circle shown in each curve of Figure 3 corresponds to the point of convergence in the iterative search for the optimal condition when the value of ϵ is equal to 0.01. In fact, the optimal value of *PeB* to minimize ϕ_{11} values ranges from about 80 to 95, depending on T in this example.

The simulation result on a heat exchanger is shown in Figure 4, in which ϕ_{12} is plotted against T_x, the parameter again being T. Clearly, an increase of T_x at constant value of T

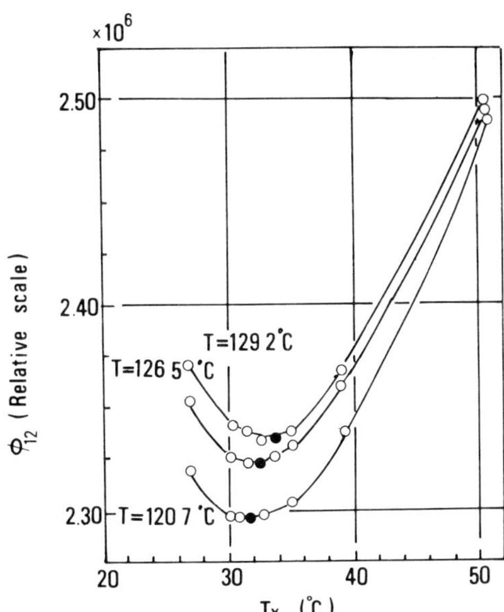

Fig. 4. Simulation of heat exchanger in continuous sterilization, provided: V = 140 m^3, t_F = 116 h, and S_B = 8 ton h^{-1}

results in an increase of A (i. e., the investment cost of the heat exchanger) according to Eq. (10), though savings of Q_W can be expected from Eq. (11). Actually, an increase of T_x of 10 °C between about 30 and 50 °C results in an increase of A to the extent of about 10%. This implies an increase of investment of about 6%; however, the fact that the value of ϕ_{12} increases by only 4% when T_x is raised from 40 to 50 °C in this example might be ascribed to the savings of Q_W.

Another observation, that a decrease of T_x below about 30 °C in Figure 4 is accompanied by the increase of ϕ_{12}, arises principally from the fact that savings of investment cost for the heat exchanger are offset by the increase of utilities cost for cooling water [cf. Eq. (11)]. Solid circles in the figure indicate the converged values of T_x for each value of T; the convergence tolerance, ϵ, employed in this instance was $\epsilon = 0.05$. The consecutive use of the golden section methods in this optimization requires another search for T, and the convergence tolerance used for ϕ_1 was $\epsilon = 0.01$.

Figure 5 demonstrates a simulation example in aeration and agitation. The increase of ϕ_2 following that of Q_A in the figure emerges since the increase of investment and utilities cost for the air compressor and air filter prevails over savings for the agitator and its utilities, whereas a decrease of Q_A causes an enhancement of P_g at a constant value of K_vp [cf. Eqs. (16)–(21)]. The decrease of Q_A results in more expenditure on agitator power, despite savings on the sterile air supply system.

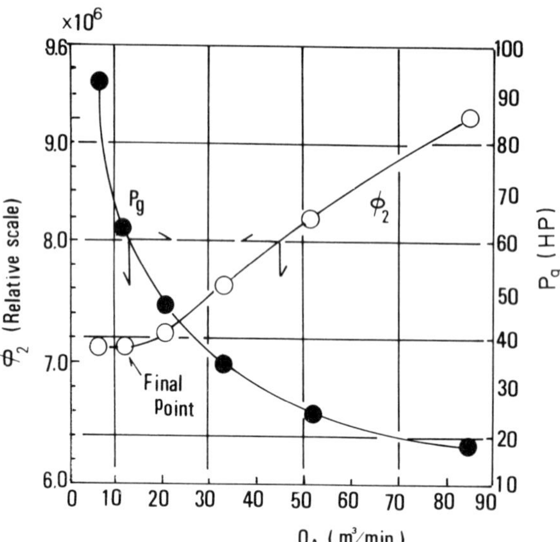

Fig. 5. Simulation of aeration and agitation of the culture, provided: $V = 140$ m³, $t_F = 116$ h, and $S_B = 8$ ton h^{-1}

The "final point" in Figure 5 represents the optimal value of Q_A; the convergence tolerance was taken as $\epsilon = 0.001$ in this iterative computation. It is interesting to have confirmed from Figures 3 to 5 that there definitely exists an optimal design for each particular part of the process, although these individual optima may not be consistent with overall optimization.

Figure 6 shows the overall optimization on the second level employing V, t_F, and S_B as coordination variables. Solid and open circles in the figure are for vertex numbers of 4 and 5, respectively, in the complex space of computation.

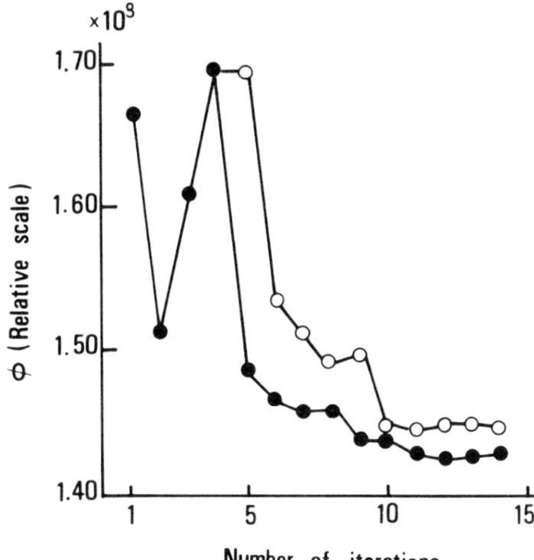

Fig. 6. Pattern of convergence in optimization. $\epsilon_1 = \epsilon_2 = 0.01$

By generating random numbers within the feasible region of constraints for V, t_F, and S_B, each vertex is commensurate with the value of ϕ, wherein ϕ_1^* and ϕ_2^* are the values optimized previously on the first level. For example, the solid circles corresponding to the iteration numbers from 1 to 4 in Figure 6 are secured after repeating the optimization on the first level; the value of ϕ for each vertex is given by $\phi = \phi_1^* + \phi_2^* + \phi_3$. Then, the modified complex method is employed and a new vertex, discarding the worst point is located by the procedure which is described elsewhere [13]. For ϵ_1 and ϵ_2 values used in this instance, see Figure 6.

Since the complex method pertains to a direct search without a mathematical formulation to guarantee that the converged value of ϕ is optimal, an indirect check of the optimum may be required. If the number of vertices is changed from 4 to 5 as a means of checking the validity of computation, the convergence pattern is different from that of the previous run, as shown in Figure 6, but the converged value of ϕ depended only slightly on the number of vertices (also see Fig. 6). Although there was still some uncertainty, the converged value of ϕ for 4 vertices (computation time $\doteqdot$ 80 s) was assumed to be sufficiently accurate in the light of the simulations represented in Figures 3–5. The values of the independent variables for the optimum were extracted from the printout and are summarized in Table 2.

By and large, the optimal values of T, PeB, T_x, and Q_A in Table 2 are not markedly different from those values expected from the preceding simulation. Nevertheless, the differences between the optimal values expected from the simulation and those optimized directly may indicate that optimization becomes more rewarding as the scale of a specific process becomes large.

Table 2. Optimized values of independent variables

Working volume of bioreactor, $V = 137.3\ m^3$
Actual reaction time, $t_F = 108$ h
Boiler capacity, $S_B = 7.9$ ton h^{-1}
Temperature of continuous sterilization, $T = 128.7\ °C$
Bodenstein number, $PeB = 88.2$
Temperature of cooling water at exit of heat exchanger, $T_X = 37.6\ °C$
Aeration rate, $Q_A = 12.3\ m^3\ min^{-1}$

It may be noted from Table 2 that the optimal value of t_F (= 108 h) in this example is far less than the time when the value of C_P levels off [cf. Eq. (32)]. At first sight it may be considered that each batch run should continue until the product concentration becomes maximal. However, the optimal design suggests that the batch should be curtailed much sooner, thus increasing the number of runs annually. This comes from taking into account various factors other than the reaction. Herein lies the significance of scale-up on an overall basis.

If a process, including the product recovery and purification, is considered, the optimization might well be conducted by using V, t_F, and S_B as variables on the first level, but the optimal value of t_F, for instance, may become different from the value in Table 2. The merit of process synthesis in operations on a large scale is self-evident [15].

Nomenclature

A	heat-transfer area of heat exchanger (m^2)
A_F	outer surface area of reaction vessel (m^2)
C_P	concentration of specific product in broth (kg m^{-3})
c	specific heat of culture medium–assumed equal to that of water (kcal $kg^{-1}\ °C^{-1}$, kcal $ton^{-1}\ °C^{-1}$)
D_i	impeller diameter (m)
D_i'	impeller diameter (cm)
D_T	diameter of reaction vessel (m)
d	inner diameter of holding pipe in continuous sterilizer (m)
E	activation energy (cal mol^{-1})
E_i	i-th utility required per batch
E_z	axial dispersion coefficient of culture medium in holding pipe ($m^2\ s^{-1}$)
e_i	unit cost of i-th utility
g_c	conversion factor (kg m $kg^{-1}\ s^{-2}$)
I_m	investment cost of m-th equipment
K_v	volumetric coefficient of oxygen transfer (kg mol 0_2 $m^{-3}\ h^{-1}\ atm^{-1}$)
k	death-rate constant of contaminants (s^{-1})
L	length of holding pipe in continuous sterilizer (m)
M_c	cost of fresh medium per batch
N	rotation speed of impeller (s^{-1})
N'	rotation speed of impeller (min^{-1})
N_C	annual number of batch runs
N_F	number of reaction vessels
N_{Re}	$d\bar{u}\rho/\mu$ = Reynolds number

Nr	$kL/\bar{u} = kt_\theta$
n	number concentration of contaminants in fresh medium at outlet of holding pipe (ml^{-1})
n_o	number concentration of contaminants in fresh medium before sterilization (ml^{-1})
P_A	annual amount required of specific product in fermented broth (kg)
PeB =	$\bar{u}L/E_z$ = Bodenstein number
P_g	power consumption of agitation in aerated bioreactor (HP)
P_o	power requirements of agitation without aeration (HP)
P'_o	power requirements of agitation without aeration (kg m s^{-1})
p	partial pressure of oxygen or average partial pressure of oxygen in bioreactor (atm)
Q_A	air flow rate (m^3 min^{-1})
Q'_A	air flow rate (cm^3 min^{-1})
Q_s	rate of steam supply in blank sterilization or to continuous sterilizer (ton h^{-1})
Q_w	rate of cooling-water supply (ton h^{-1})
R	gas constant, 1.98 cal mol^{-1} K^{-1}
S_B	boiler capacity (ton h^{-1})
T	temperature of bioreactor in blank sterilization or temperature of continuous sterilization (°C)
T_c	temperature of cooling water at inlet of heat exchanger (°C)
T_F	culture-medium temperature at exit of heat exchanger (°C)
T_o	room temperature (°C)
T_s	steam temperature (°C)
T_x	temperature of cooling water at exit of heat exchanger (°C)
t_A	annual working time (h)
t_C	cycle time for each batch (h)
t_F	time of actual reaction (h)
t_{01}	time needed for cleaning bioreactor (h)
t_{02}	time required for blank sterilization (h)
t_{021}	time required for raising temperature of bioreactor from $T = T_o$ to $T = T$ in blank sterilization (h)
t_{022}	period of time during which temperature of bioreactor is kept at $T = T$ in blank sterilization (h)
t_{023}	time required for lowering temperature of bioreactor from $T = T$ to $T = T_F$ (h)
t_{03}	time for continuous sterilization of culture medium (h)
t_{04}	time needed for inoculation (h)
t_{05}	time needed for discharging broth (h)
t_θ	mean residence time of culture medium in holding pipe (s)
U	overall heat-transfer coefficient for heat exchanger (kcal m^{-2} h^{-1} $°C^{-1}$)
U_F	overall heat-transfer coefficient for vessel wall due to natural convection (kcal m^{-2} h^{-1} $°C^{-1}$)
$\bar{u}$	mean linear velocity of culture medium passing through holding pipe (m s^{-1})
V	volume of culture medium in batch culture (= working volume of bioreactor) (m^3)
v_s	superficial air velocity based on empty cross-sectional area of vessel (m h^{-1})
W	heat capacity of bioreactor (kcal $°C^{-1}$)
x	axial distance in holding pipe
α	fraction of steam condensed in blank sterilization
α'	empirical constant (min^{-1})
α_L	depreciation (= 0.3) multiplied by Lang factor (= 3.6)
β'	empirical constant (= E/R) (K)
ϵ	convergence tolerance
ϵ_1	convergence tolerance of first stopping criterion of modified complex method
ϵ_2	convergence tolerance of second stopping criterion of modified complex method
ζ =	$(1 + 4\,Nr/PeB)^{1/2}$
λ	latent heat of vaporization of water (kcal ton^{-1})
μ	viscosity of culture medium at sterilization temperature (kg m^{-1} s^{-1})
ρ	density of culture medium (ton m^{-3}, kg m^{-3})

ϕ objective function of process (cost/year)
ϕ_1 objective function for continuous sterilization of culture medium, holding pipe and heat exchanger included (cost/year)
ϕ_2 objective function of aeration and agitation and bioreactor (cost/year)
ϕ_3 objective function for adjunct equipment items and utilities (cost/year)
ϕ_{11} objective function for holding pipe in continuous sterilization (cost/year)
ϕ_{12} objective function for heat exchanger in continuous sterilization (cost/year)

Subscripts
1, 2, 3 continuous sterilization of culture medium, aeration and agitation, and adjunct equipment items, respectively
i electricity, steam, water, or air in utilities
i' 2 or 3
m m-th equipment item

Superscript
* optimized value for individual stage

References

1. Aiba, S., Humphrey, A. E., Millis, N. F.: In: Biochemical Engineering, 2nd Edition. New York: Academic Press 1973.
2. Aiba, S., Okabe, M., Okada, M.: Progress Report No. 79, Biochem. Eng. Lab., Inst. Appl. Microbiol., University of Tokyo, December, 1972.
3. Aiba, S., Okabe, M., Okada, M.: Progress Report No. 82, Biochem. Eng. Lab., Inst. Appl. Microbiol., University of Tokyo, 1973.
4. Bartholomew, W. H.: Adv. Appl. Microbiol. **2**, 289 (1960).
5. Cooper, C. M., Fernstrom, G. A., Miller, S. A.: Industrial and Engineering Chem. **36**, 504 (1944).
6. In: Cost and Optimization Engineering, F. C. Jelen (ed.). New York: McGraw-Hill 1970, p. 317.
7. Fukuda, H., Sumino, Y., Kanzaki, T.: J. Ferment. Technol. **46**, 838 (1968).
8. Humphrey, A. E., Reilly, P. J.: Biotechnol. Bioeng. **7**, 229 (1965).
9. Koga, S., Bury, C. R., Humphrey, A. E.: Appl. Microbiol. **15**, 683 (1967).
10. Kono, T., Asai, T.: Biotechnol. Bioeng. **11**, 19 (1969).
11. Levenspiel, O.: Industrial and Engineering Chem. **50**, 343 (1958).
12. Maxon, W. D.: J. Biochem. Microbiol. Tech. & Eng. **1**, 311 (1959).
13. Okabe, M., Aiba, S., Okada, M.: J. Ferment. Technol. **51**, 594 (1973).
14. Okabe, M., Aiba, S.: J. Ferment. Technol. **52**, 279 (1974).
15. Okabe, M., Aiba, S.: J. Ferment. Technol. **53**, 730 (1975).
16. Oldshue, S. Y.: Biotechnol. Bioeng. **8**, 3 (1966).
17. In: Process Systems Engineering, T. Takamatsu (ed.). Tokyo: Nikkan-Kogyo 1972, p. 104.
18. Taguchi, H., Imanaka, T., Teramoto, S., Takatsu, M., Sato, M.: J. Ferment. Technol. **46**, 823 (1968).
19. Terui, G., Okazaki, M., Kinoshita, S.: J. Ferment. Technol. **45**, 497 (1967).
20. Toda, K.: Ph. D. Thesis, Osaka Univ., 1967.

The Redox Potential: Its Use and Control in Biotechnology

L. Kjaergaard
Department of Applied Biochemistry, The Technical University of Denmark,
Block 223, DK-2800 Lyngby, Denmark

Contents

This review describes the theory for and the measurement of the redox potential and the use and control of redox potential in biotechnology.
The theory for the redox potential from a thermodynamic point of view is outlined and the reasons for that the measurement of the redox potential is only an indication of the oxidative status in a complex medium as e.g. growth media in which several reactions take place is mentioned.
The works concerning redox potential in microbial and various biochemical systems and in soil science that has been published during the last five years are reviewed and the possibility of establishing biochemical fuel cells is discussed. Furthermore the few works concerning regulation by means of the redox potential are reviewed. It is concluded that although the reactions that determine the redox potential in biochemical systems are complex and almost unknown the redox potential may be a parameter which can give valuable information about the metabolism taking place in microbial cultures.

1. Introduction

Many aerobic microbial processes take place at concentrations of dissolved oxygen that are impossible to measure using commercial probes for dissolved oxygen. However, the supply of oxygen to the cultures is essential to the microorganisms. Since the range of the oxygen-limited state is infinite, i.e., from the almost completely anaerobic state through the so-called partially aerobic state (Pirt, 1957) to a state where oxygen can be measured by means of a commercial probe (i.e., above 10 μM O_2), it is important to have a tool that can provide information about the degree of oxygen limitation. Such a tool is the measurement of the redox potential in the growth liquid. In this paper the significance and some of the applications of the redox potential, e.g., as a parameter that can be used for regulating purposes, will be outlined.

2. Theory of Redox Potential

In chemistry the definition of the reduction-oxidation potential is related to a pair of components, of which one (Ox) can be reduced reversibly to the other (Red) by means of moving n electrons (e):

$$\mathrm{Ox} + ne^- \rightleftharpoons \mathrm{Red} \qquad (1)$$

and the redox potential E_h for this equilibrium can be calculated using the Nernst equation:

$$E_\mathrm{h} = E^\mathrm{o} + \frac{RT}{nF} \ln \frac{\text{activity of Ox}}{\text{activity of Red}}$$

$$= E^\mathrm{o} + \frac{1}{n} \cdot \frac{RT}{F} \cdot 2.303 \cdot \log_{10} \frac{\text{activity of Ox}}{\text{activity of Red}}, \qquad (2)$$

where E_h is the potential referred to the normal hydrogen electrode, E° is the standard potential of the system at 25 °C when the activities of all reactants are unity, R is the gas constant, T is the absolute temperature, F is the Faraday constant, and

$\frac{RT}{F} \cdot 2.303 = 59.1$ mV.

Since most biological liquids are extremely complex and since normally a variety of numerous, simultaneous, successive, and parallel chemical and biological reactions, many of which are irreversible, take place, many authors consider the redox potential in a biological liquid as a measurement of little value (Harrison, 1972).

The measured redox potential is related to all the pairs of reducible and oxidizable compounds found in the medium. In the same manner in which the pH value (the negative logarithm of the activity of the hydrogen ion, which can react with a variety of reactants in a liquid) provides information about hydrogen ion activity (though the pH is measured as a potential), the redox potential provides information about the "activity of the electrons" in the medium. Normally the pH value in a medium varies due to the growth of the microorganisms and the concomitant catabolism of the substrate. This variation can be suppressed by adding high-buffer capacity components to the medium. Lack of buffers gives an exact pH variation that provides information about the change in activity of hydrogen ions due to the growth and the catabolism of the substrate.

The same is valid for the redox potential. When only one pair of Red-Ox compounds is present in the medium, the measured redox potential indicates immediately which equilibrium has been obtained in reaction (1).
Normally there will be several redox pairs so that some Red compounds are oxidized to the Ox compounds.
During microbial growth the most interesting redox systems are the systems in which O_2 takes part. The redox reactions taking place during microbial growth can be formulated as

$$\sum_{i=1}^{n_1} \mathrm{Ox}_{1i} + b_1 e^- \rightleftharpoons \sum_{i=1}^{n_1} \mathrm{Red}_{1i}, \tag{3a}$$

$$\sum_{j=1}^{n_2} \mathrm{Red}_{2j} \rightleftharpoons \sum_{j=1}^{n_2} \mathrm{Ox}_{2j} + b_2 e^-, \tag{3b}$$

where b_1 and b_2 are the numbers of electrons needed for reduction of $\sum_{i=1}^{n_1} \mathrm{Ox}_{1i}$ and $\sum_{j=1}^{n_2} \mathrm{Ox}_{2j}$, respectively, and

$$\frac{b_2 - b_1}{4} O_2 + (b_2 - b_1)e^- + (b_2 - b_1)H^+ \rightleftharpoons \frac{b_2 - b_1}{2} H_2O. \tag{3c}$$

Addition of Eqs. (3a), (3b), and (3c) gives

$$\frac{b_2 - b_1}{4} O_2 + (b_2 - b_1) H^+ + \sum_{i=1}^{n_1} \mathrm{Ox}_{1i} + \sum_{j=1}^{n_2} \mathrm{Red}_{2j} \rightleftharpoons$$

$$\frac{b_2 - b_1}{2} H_2O + \sum_{i=1}^{n_1} \mathrm{Red}_{1i} + \sum_{j=1}^{n_2} \mathrm{Ox}_{2j}. \tag{3}$$

For the three reaction systems (3a), (3b), and (3c), the redox potential can be calculated using the Nernst equation:

$$E_h = E_1^\circ + \frac{RT}{b_1 F} \cdot \ln \frac{a_{\sum_{i=1}^{n_1} \mathrm{Ox}_{1i}}}{a_{\sum_{j=1}^{n_1} \mathrm{Red}_{1i}}}, \tag{4a}$$

$$= E_2^\circ + \frac{RT}{b_2 F} \cdot \ln \frac{a_{\sum_{j=1}^{n_2} \mathrm{Ox}_{2j}}}{a_{\sum_{j=2}^{n_2} \mathrm{Red}_{2j}}}, \tag{4b}$$

$$= E^\circ_{O_2/H_2O} + \frac{RT}{4F} \ln \frac{a_{O_2} \cdot a^4_{H^+}}{a^2_{H_2O}} \tag{4c}$$

from which it can be seen that

$$E_h = \frac{1}{3}\left[E_1^\circ + E_2^\circ + E^\circ_{O_2/H_2O} + \frac{RT}{F}\ln\left(\frac{a_{\sum_{i=1}^{n_1} Ox_{1i}}}{a_{\sum_{i=1}^{n_1} Red_{1i}}}\right)^{\frac{1}{b_1}} \cdot \left(\frac{a_{\sum_{j=1}^{n_2} Ox_{2j}}}{a_{\sum_{j=1}^{n_2} Red_{2j}}}\right)^{\frac{1}{b_2}} \cdot \left(\frac{a_{O_2}\cdot a^4_{H^+}}{a^2_{H_2O}}\right)^{\frac{1}{4}}\right]. \quad (4)$$

This will be the redox potential in the aerobiosis if an equilibrium is obtained. Because Eqs. (4a), (4b), and (4c) are valid in the equilibrium, it is seen by rearrangement of (4c) that

$$E_h = E^\circ_{O_2/H_2O} + \frac{RT}{4F}\ \ln a_{O_2} - \frac{RT}{F}\cdot 2.303 \cdot \text{pH} \quad (5)$$

as $a_{H_2O} = 1$ and pH = $-\log_{10} a_{H^+}$. The standard potential $E^\circ_{O_2/H_2O} = 1223$ mV. From (5) it can be seen that for an equilibrium the E_h will decrease by 14.8 mV for a decrease in a_{O_2} by a factor of 10. This is in opposition to what has been reported by several authors (Schuldiner *et al.*, 1966; Jacob, 1971; Kjaergaard, 1976). However, the latter two refer to Schuldiner *et al.* and obviously they have not calculated the value themselves. Neither have Schuldiner *et al.* (1966) calculated the value 59.1 mV; however, by measuring the open-circuit rest potential on bright platinum as a function of the partial pressure of oxygen in an electrolyte consisting of 1 M H_2SO_4 maintained at 25° ± 2 °C when the partial pressure of O_2, $\mathbf{p}_{O_2}$, was in the range 10^{-6}–10^{-2} atm, they found that a 10-fold increase in $\mathbf{p}_{O_2}$ yielded an increase in the measured potential of 0.06 V. In the same paper, they showed that the E_h value decreases by 0.030 V when the partial pressure of H_2, p_{H_2}, increases by a factor of 10 for p_{H_2} in the range 10^{-6}–10^{-2} atm.

Since the E_h for the hydrogen–hydrogen ion redox system

$$2\,H^+ + 2\,e^- \rightleftharpoons H_2 \quad (6)$$

is

$$E_h = E^\circ_{H^+/H_2} + \frac{RT}{2F}\ln\frac{a^2_{H^+}}{a_{H_2}}$$

$$= -\frac{RT}{2F}\ln a_{H_2} - \frac{RT}{F}2.303\ \text{pH} \quad (7)$$

in which $E^\circ_{H^+/H_2}$ is defined as zero, it can be calculated that a decrease in E_h of 29.6 mV corresponds to a 10-fold increase in p_{H_2}, as found by Schuldiner *et al.*

However, other authors (Ishizaki *et al.*, 1974) calculated the decrease in E_h for a 10-fold decrease in DO (concentration of dissolved oxygen) to be 14.8 mV. As early as 1941, Joergensen (1941) described these equations and introduced a concept analogue to the pH, namely, the rH, which is defined as

$$\text{rH} = -\log_{10} a_{H_2}. \quad (8)$$

Combination of (7) and (8) gives

$$rH = 2(pH + \frac{F}{RT} \cdot \frac{1}{2.303} \cdot E_h). \tag{9}$$

This rH value can vary from below rH = 0, corresponding to a solution in which p_{H_2} = 1 atm and pH = 0, to above rH = 42, corresponding to a solution in which p_{O_2} = 1 atm and pH = 0.

Although this rH value may be more informative than the E_h value, as the oxidative status is immediately clear from the rH value, it is seldom used.

In addition to introducing the rH concept, Joergensen (1941) mentioned that biological liquids are not often in equilibrium and that therefore the measured E_h or rH values are not correct from a thermodynamic point of view. However, Joergensen concluded that the values obtained will yield information about the oxidative status of the liquid.

More than 20 years earlier Gillespie (1920), who measured redox potentials in cultures of bacteria, had reached the same conclusion: "Biological significance may be capable of demonstration even if chemical significance may not."

Another reason why the measured redox potential is not correct in a thermodynamic sense is that most of the systems used to investigate microbial growth and metabolism are open systems in which no equilibrium can be obtained.

Despite these facts, many scientists have used the E_h measurements, and many have succeeded in finding correlations between the E_h value and other parameters, extra-cellular as well as intracellular. However, as late as 1971, it was not possible to obtain a better understanding of the theory of the redox potential. Therefore Jacob (1971) proposed that the E_h value should not be called the redox potential but the platinum electrode potential when measured by a platinum electrode. In this paper the potential will be called the redox potential or the E_h value.

3. Measurement of Redox Potential

In principle there are two ways of measuring a redox potential: (1) by redox dyes and (2) by electrodes.

Measurement of the redox potential by dyes is not exact and requires a number of different dyes to obtain just a semiquantitative measurement; furthermore, many of these dyes may be toxic to the cells or may inhibit the enzyme activities in biological liquids (Hill, 1973). Therefore, this method is not used in biochemical engineering except in classification assays.

3.1 Electrodes

E_h can be measured by an electrode because the potential of an electrode immersed in a liquid is dependent on the activity of the electrons in the liquid. Higher electron activity which will tend to force reaction (3) toward the right, will decrease the potential in accordance with Eq. (4).

The electrodes to be used for the measurements must fulfill a number of requirements as outlined below. Kantere (1972) discussed the requirements that the electrodes must meet and gave 3 general technical requirements:

1. High electronic conductance of the electrode material.
2. Neutrality toward the analyzed liquid, i.e., absence of redox interactions between the electrode material and the solution.
3. Sufficient speed of electronic exchange at the electrode-solution boundary of the redox system (or a mixture of these systems) that would guarantee stable values of the electrode potential and reversibility to the specific redox system.

The first two of these requirements are idealized and can not be realized completely with any type of electrode.

However, 6 types of electrodes are found to meet the requirements to the extent that they are used in redoximetry:

1. Smooth, massive electrodes of noble metals: platinum, gold, or less often rhodium or iridium.
2. Thin-layer electrodes of the preceding noble metals with enlarged surfaces.
3. Electrodes of mercury or other metals.
4. Electrodes of carbon varieties, preferably graphite.
5. Electrodes made of semiconductors (stannic oxide, titanic oxide), and
6. Glass electrodes.

Of these 6 types, the first two are especially suitable for redox measurements. Type No. 6 is ideal for measuring pH values (Linnet, 1970) but may be used for redox potential measurements, too. Electrodes of the first two types have been thoroughly investigated by Jacob (1971), who in an excellent paper investigated the influence of several physical parameters on the measured E_h value. Jacob found that the electrode should have a certain size (not less than 2 mm^2) and that it was important to clean the electrode thoroughly before use [mechanical cleaning followed by polishing with cerium oxide was best (Jacob, 1974)].

3.2 Reference Electrodes

Another important factor for measuring the redox potential is the reference electrode. In principle this should be the hydrogen electrode whose own potential is defined as zero. However, it is more convenient to use one of the more common reference electrodes, e.g., a calomel electrode or more often a silver/silver chloride electrode. The potential that is measured between a reference electrode and the redox electrode must then be corrected for the potential of the reference electrode. At 37 °C the potentials of a silver/silver chloride electrode and a calomel electrode are 264 mV and 214 mV, respectively (Linnet, 1970). The redox potentials mentioned in this paper are corrected for the potential of the reference electrode. When pH measurements are also performed in the liquid in which the redox potential is to be measured, the same reference electrode can be used for both measurements.

3.3 Calibration

A great problem with redox potential measurements is the calibration of the electrodes.

As was mentioned in Section 2 there will be no equilibrium in a liquid growth medium, and therefore the measured potentials are not exact in a thermodynamic sense. However, the only correct method for adjusting the electrodes is to measure the potential of a well-defined solution in which an equilibrium exists. Jacob (1971) recommends calibration in a suspension of a 16-h *Proteus vulgaris* culture in which the potential reaches its most negative value.

It must be emphasized that the E_h value that can be obtained using *P. vulgaris* is not the lowest reported. *Desulfovibrio sp.* are able to grow at redox potentials as low as −550 mV (Zobell, 1975).

The method described by Jacob (1971) is also recommended by Zwarun (1975), who used redox measurements in media for anaerobic blood culture, where the degree of reduction is important.

A better method, which is used in our laboratory, is to register the potentials obtained in two saturated solutions of quinhydrone at two different pH values. With this method the sensibility of the redox electrode to pH is measured and the absolute value is registered. At 25 °C the potential of the quinhydrone electrode is (Linnet, 1970):

$$E_{h,\ \mathrm{quinhydrone}} = 699 - 59.1\ \mathrm{pH}. \tag{10}$$

This calibration is performed before and after every use of the electrodes. If a deviation from (10) is observed, the electrode is cleaned. Furthermore, the redox potential of the freshly prepared medium saturated with oxygen gives the same E_h value in every preparation.

This calibration method is considered to be better than that mentioned by Jacob (1971) since the measurement with quinhydrone is performed in a system in which an equilibrium exists. A good calibration method is important, especially for medical use, where only small changes in the E_h value may be observed (Hays and Mandell, 1974).

3.4 Electrical Instrumentation

Since the E_h measurement is a potential measurement, it should be possible to read the value directly from a millivoltmeter. However, in order not to disturb the measurement the signal should be amplified as for the pH measurement. In fact general pH instruments can be used for the amplification. The important thing is to use a high impedant instrument as amplifier so that only a small current is drained from the system. Normally the amplifier converts the signal to a standard current (e.g. 0–20 mA) so that the redox potential can be logged together with other parameters.

4. Significance of Redox Potential

Despite the fact that the redox potentials measured in biological liquids are not thermodynamically correct, there have been a number of investigations of the influence of redox potential on several biological systems during the past years. Only a few of these deal with microbial systems. Almost all the authors find that the redox potential is a valuable measurement when the concentration of dissolved oxygen cannot be measured.

4.1 Redox Potential in Microbial Growth Systems

Wimpenny (1969) showed that there was a connection between the E_h value of a culture of *Escherichia coli* and the levels of the activities of three tricarboxylic acid-cycle enzymes. The different levels of E_h were established by changing the aeration. Later Wimpenny and Necklen (1971) investigated the different physiologic phases that can be identified during the transition from anaerobiosis to aerobiosis for *E. coli* and *Klebsiella aerogenes.*

They found that the E_h value increased for every change in physiologic phase toward a higher degree of aerobiosis, and in fact they found that the levels of tricarboxylic acid enzymes, cytochromes, steady-state ATP pool, growth yield, and levels of hydrogenase were dependent on the E_h value. However, they concluded that the redox potential was not clearly understood due to the reasons already mentioned in Section 2. Five years later Wimpenny (1976) proposed that the redox potential is only a measure of the oxidation of a few compounds (electron donors), namely, those compounds whose oxidation takes place with a relatively high velocity due to the presence of catalysts (enzymes) specific for the oxidation. The potential of these few reactions then swamps all other potentials.

This theory is not in opposition to the information presented in Section 2 for the E_h value since Wimpenny (1976) also realized that there is no steady state in microbial cultures.

Since all microbial processes are controlled by the metabolism of the cells, it is probable that the production rate depends on the E_h value. Sukharevich *et al.* (1970) found that the redox potential influenced the biosynthesis of the antibiotics levorin A and B, which are produced simulaneously by *Actinomyces levoris,* but with the A compound preferably produced at high values of E_h and the B compound preferably produced at low redox potentials. The methods used for maintaining the redox potential at a fixed value were partly physical (changing of aeration rate) and partly chemical (addition of chemicals); therefore, the observed effect may be due to the different methods instead of the different redox potentials. Whether the change in antibiotic production could be correlated to metabolic changes of the microorganisms was not investigated.

Kantere (1972) investigated how the redox potential, measured with different types of electrodes, changed during the growth of *Candida utilis* at constant pH. A decrease in redox potential was common for all the electrodes, although the magnitude of the change was dependent on the type of electrode. When the concentration of dissolved oxygen was brought to the initial level, the redox potentials of all the electrodes reacquired the initial values within ± 5 mV. Furthermore, Kantere (1972) tried to determine the amount of reductants produced during growth by titrating the medium, a 24-h culture and a 48-h culture, with 0.001 *N* $K_3Fe(CN)_6$. These volumetric redox titrations gave titration curves of the normal S-shape, but the redox potentials at which the inflexion point (equivalence point) was found were different for the three titrations. The results of these titrations are summarized in Table 1.

These results seem to show that some reductants are produced during growth. Kantere concluded from these results that the redox buffer capacity of the culture has been increased during growth. If the capacity has been changed only because of a change in

Table 1. Volumetric redox titration of medium for and cultures of *Candida utilis* with 0.001 *N* $K_3Fe(CN)_6$ (Kantere, 1972).

	Inflexion point mV	Concentration of reductant *M*
Culture medium[a]	165	$6.5 \cdot 10^{-4}$
24-h culture	130	$20.0 \cdot 10^{-4}$
48-h culture	70	$66.0 \cdot 10^{-4}$

[a] The composition of the medium (g/l) was as follows: KNO_3, 6.8; $MgSO_4$, 0.2; K_2HPO_4, 0.1, and glycerine 25; pH was 4.5

the concentration of the initial buffer system, the inflexion point, i.e. the average between the individual standard potentials (Hägg, 1965), should not change. Therefore, it must be concluded that the composition of the reductants has changed during growth. It is not evident how the redox potential could be brought to the initial value by saturation with oxygen. It must be assumed that this is not possible until the produced reductants in the medium have been oxidized, unless the redox potential only reflects the activity of oxygen and not the concentration of reductants that might not be oxidizable with oxygen.

Later Kantere, together with Balakireva and Robotnova (Balakireva *et al.,* 1974) defined a buffer capacity π as

$$\pi = \frac{nF}{RT} \frac{1}{2.303} \cdot \frac{1}{C} \cdot C_{red} (C - C_{red}), \qquad (11)$$

where C is the sum of concentrations of the oxidants, C_{Ox}, and reductants, C_{red}. They claimed that solutions containing chemically irreversible systems and redox systems with weak concentrations of oxidants and reductants are not buffered or otherwise characterized by a small π value. In such solutions, the electron activity, which they consider as a measure of the oxidizing (reducing) ability, is not stable. Microbial cultures are examples of such solutions. In the same article (Balakireva *et al.*, 1974) it is supposed that there are only comparatively low concentrations of oxidants and reductants in microbial cultures and that a considerable shift of the a_{ox}/a_{red} ratio to the reduced form is typical for many media and causes a small buffer capacity of such cultures.

Furthermore, Balakireva *et al.* (1974) considered it to be a characterization of a living system (e.g., a microbial culture) that there is no thermodynamic steady equilibrium state. The measured E_h value that may be constant in a system is called a stationary potential and is considered a purely instrumental notion.

From these considerations it seems plausible that the stationary potential in a microbial culture can reflect the a_{ox}/a_{red} ratio of a redox system present in higher concentration than other redox systems. In other words it seems plausible that the redox potential is determined by one redox systems when the buffer capacity is low.

Here an analogy may be drawn to pH measurement. In a microbial culture that runs at pH = 4 the presence of phosphate suppresses the pH-change due to production of organic acids much less than would be the case at pH = 7.

Kantere (1972) found that the redox potential that was measured during the aerobic growth of *C. utilis* reflected the concentration of dissolved oxygen.
The titrations performed with the cultures showed that the amount of reductants in the culture increased for older cultures and that the lower value at which the redox buffer capacity was highest decreased for older cultures and was always below 50–100 mV, indicating that the buffer system changes during the growth of the culture.
Balakireva *et al.* (1974) also showed that the stationary potential was 'more stationary' in microbial cultures than in the pure medium. They carried out this investigation by observing the transients of the stationary potential after anode and cathode polarization of the electrode, and they found (Table 2) that the potential returned to the original value most rapidly in the complete culture, more slowly in the supernatant, and very slowly in the pure medium.

Table 2. Transients for redox potential after anode and cathode polarization (the initial deviation was approximately 200 mV) (Balakireva *et al.*, 1974)

	Time after polarization min	Deviation in mV in		
		Culture	Supernatant	Medium
Anode polarization	20	50	130	165
	60	0	55	110
	300	0	30	85
Cathode polarization	20	20	105	145
	60	0	50	115
	300	0	45	90

From the results in Table 2 Balakireva *et al.* (1974) concluded that the electrode potential is not accidental but rather a stationary potential that reflects the redox conditions in the medium. Furthermore, it is obvious that the microorganisms as well as the products excreted from them influence the measured E_h value.
The preceding authors all seem to be satisfied with the possibility for measuring the redox potential, and they try to relate the redox potential to the presence of some redox systems, probably oxygen-water. However, they do not compare the measured E_h values with the concentration of dissolved oxygen to elucidate whether oxygen is the acting redox oxidant and water the reductant.
Kantere (1972) and Balakireva *et al.* (1974) did not compare the obtained results for E_h and dissolved oxygen (DO), neither did Jacob (1971), who refers to Schuldiner *et al.* (1966). Wimpenny (1969) and Wimpenny and Necklen (1971) also might have compared the E_h values with the DO values in the chemostat experiments. In Table 3 some of the measured E_h values are compared to the E_h values that can be calculated from the DO values and the E_h value corresponding to the highest DO value in the series.
Although these calculations might be very defective, it is significant that the redox potential at lower DO-values decreases much more than that which can be calculated

Table 3. Comparison between measured and calculated E_h values (E_h measured with a Pt electrode)

DO atm	E_h measured mV	E_h calculated[a] mV	Ref.
C. utilis			
0.17	125	–	Balakireva *et al.* (1974)
0.12	65	123	Balakireva *et al.* (1974)
0.04	15	116	Balakireva *et al.* (1974)
E. coli			
0.15	300	–	Wimpenny (1969)
0.06	300	295	Wimpenny (1969)
0.03	230	289	Wimpenny (1969)
0.015	120	285	Wimpenny (1969)
K. aerogenes			
0.18	340	–	Wimpenny and Necklen (1971)
0.08	320	335	Wimpenny and Necklen (1971)
0.05	310	332	Wimpenny and Necklen (1971)
0.04	300	330	Wimpenny and Necklen (1971)
0.01	290	332	Wimpenny and Necklen (1971)
S. aureus			
0.19	340	–	Jacob (1970)
0.15	330	338	Jacob (1970)
0.11	320	336	Jacob (1970)
0.05	295	330	Jacob (1970)

[a] The calculation of E_h is based upon Eq. (5)

from the Nernst equation. This fact, together with the observations of Balakireva *et al.* (1974) that the redox potential after a polarization returns to the original value much faster in a complete culture than in the supernatant of the culture or the pure medium, seems to show that the redox potential is not only a measure for the concentration of dissolved oxygen.

In his review on growth, oxygen, and respiration, Harrison (1973) argued that oxygen is merely a part of the redox environment of the cell and influences the cell passively as an electron sink. However, the conclusion of Harrison that the concept of overall redox potential is of little value in studies of growing microbial cultures, seems a little premature, as shown in the next section.

Having established that the redox potential in microbial cultures is a concept that yields information about the oxidative status in the culture, it should be appropriate to look at the effects of the redox potential upon the cell mass produced and the metabolism. Wimpenny's investigations have been mentioned already. Shibai *et al.* (1974) investigated the influence of redox potential upon the microbial production of inosine. In a previous paper (Ishizaki *et al.*, 1974) the same authors had demonstrated that the redox potential, measured with a platinum electrode, depended on pH value, dissolved oxygen concentration, equilibrium constant, and oxidation reduction potentials in the

liquid. They found that the decrease in E_h in response to a tenfold decrease in DO, which according to the Nernst equation ought to be 14.8 mV, varied from 15 mV to 100 mV depending on the medium in which the measurement was made.
By changing the rate of air flow and the agitation speed, the batchwise production of inosine by *Bacillus subtilis* was performed with different E_h values in the stationary phase. Shibai *et al.* found, in good agreement with Wimpenny and Necklen (1971), that the metabolism was strongly influenced by the redox potential, leading to production of different by-products, depending on the E_h value in the stationary phase: thus, there was no accumulation of inosine in the medium when $E_h \leqslant -160$ mV, but the accumulation of lactic acid in the culture was increased. Furthermore, they investigated the effect of chemicals that inhibit cell respiration. It was found that addition of cyanide, azide, or 2,4-dinitrophenol in concentrations that completely inhibited respiration caused an increase in DO but a decrease in E_h, indicating that the E_h is not only a measurement of the DO.
A similar observation was made by Andreeva (1974), who investigated the physiologic and biochemical changes occurring in the yeast *C. utilis* when the redox potential was changed by addition of either ascorbic acid or potassium hexacyanoferrate (III). In these experiments the pH as well as the DO was maintained at constant levels. In the steady state without addition of redox compounds, the redox potential was 220 mV. A change of 55 mV by automatic titration with either hexacyanoferrate (III) (for E_h = 275 mV) or ascorbic acid (for E_h = 165 mV) led to a drastic decrease in the cell yield, an increase in consumption of the carbon source (glycerol) and of the N source (NH_4^+). It is not indicated in the paper whether the added amount of redox compounds caused any decrease in the maximum growth rate, which might explain some of the observed changes. However, it must have been assumed that the maximum growth rate was unaffected and therefore the degree of wash out in the chemostat culture was the same in all experiments, and hence the observed changes were true. Neither an increase in the cellular content of protein nor of polysaccharides could explain the increase in consumption of the C- and N sources. An increase in accumulation of volatile acids, pyruvic acid, and α-ketoglutaric acid was observed but the increase, on a weight basis, was much smaller than the observed additional consumption of the C source. The consumption of phosphorus was almost unaffected by the changes in redox potential, but due to the decrease in cell concentration the content of phosphorus in the cells increased by a factor of 2; especially the concentrations of polyphosphates were changed when the redox potential was changed. Andreeva interpreted these significant changes as an uncoupling of the energy production processes. These results are consistent with the results obtained by Wimpenny and Necklen (1971) who found that most of the tricarboxylic acid-cycle enzymes were produced in maximum quantities for $E_h \sim 200$ mV —the enzymes that are essential for the production of energy by oxidative phosphorylation.
From these studies it seems that the redox potential provides information about more than the oxidative status of the culture. It is obvious that any change in the redox potential from the per se established value will lead to a decrease in effectivity of the cell. However, it must be emphasized that the extracellular redox potential is different from the redox potentials that are measured in cell-free extracts and that are closely

related to the redox potentials for the different steps in the respiratory chain (Zs.-Nagy and Ermini, 1972).

Kjaergaard (1976) showed recently that the ineffective by-production of acetic acid by *Bacillus licheniformis* depended heavily on the redox potential. Unlike in most of the preceding studies the redox potential was changed by changing something other than the aeration and addition of redox chemicals. Using the fact that oxygen consumption is dependent on the concentration of microorganisms in a chemostat, the oxygen demand was increased by increasing the concentration of the growth-limiting component in the medium, which increased the cell concentration. In this type of experiment and in experiments with varied aeration, the acetic acid production was heavily dependent on the redox potential. These results are consistent with the results reported above.

In a chemostatic study of the anaerobe *Bacteroides fragilis* Onderdonk *et al.* (1976) found that a change in E_h from the steady-state value –75 mV to +300 mV did not influence the cell yield, when the change was due to addition of hexacyanoferrate(III). When the redox potential was increased to +250 mV by aeration of the culture, the growth ceased completely. The first results might seem to be inconsistent with the results obtained by Andreeva (1974), but it must be emphasized that Onderdonk *et al.* used an anaerobic organism for which oxygen may be toxic, whereas Andreeva used an aerobic organism. However, it must be concluded from the preceding experiments that the redox potential reflects several redox systems of which oxygen is only the oxidant in one system. From the above-cited results it seems reasonable to recommend the use of the redox potential for regulatory purposes, especially if the characteristics of the redox potential are not generalized from one organism to another. On the contrary, the redox potential may also be used as a tool in the identification technique (Shikova *et al.,* 1972; Blagova and Belozerova, 1970).

From the viewpoint of an engineer, it should, as for most of the published experiments in biotechnology, be extremely interesting to know whether the reported observations are also valid for processes taking place in more complex substrates. There are few experiments with complex substrates in the literature, and only Shibai *et al.* (1974) investigated the redox potential in such complex media. In our laboratory we have started a series of experiments in which complex media are used for production of α-amylase by *B. licheniformis* and in which the influence of the redox potential on metabolism and enzyme production will be investigated. Thus far the results seem to indicate that the redox potential for these complex media has the same time course as that for simple salt media.

4.2 Redox Potential in Biochemical Systems

From general biochemistry it is well known that several of the metabolic reactions are redox reactions. The most thoroughly investigated reactions are those in the respiratory chain, and quantitative calculations of the possible energy production are even based on the redox potential (Zs.-Nagy and Ermini, 1972). Furthermore, relations between the redox potential and photosynthesis have been investigated (Einor, 1973; Cogdell *et al.*, (1973). The measurements and interpretations of the redox potentials in such

systems (chloroplasts, chromatopheres, etc.) are often further complicated by the presence of membrane potentials (Cogdell *et al.*, 1973).
Analogous to the effects of redox potential on the growth of *Candida utilis* (Andreeva, 1974) an effect on pure enzymatic reactions was shown by Gemant (1974), who investigated the influence of the redox potential on the activity of polyphenol oxidase. He found that a lowering of the redox potential obtained by addition of ascorbic acid decreased the rate of the oxidative enzyme reaction. This finding may also be due to inhibition by ascorbic acid.

4.3 Redox Potential in Soil Science

While oxygen normally is the most important electron acceptor in aerobic systems, there are other important electron acceptors in anaerobic systems. Also in more complex systems such as soils there are electron acceptors other than oxygen.
This means that these electron acceptors may play an important role in the redox potential although they do not produce any redox potential themselves. Wimpenny and Necklen (1971) mentioned that nitrate and possibly nitrite might alter the E_h value of a culture through the cell. In soil and in waste water, nitrate and nitrite are electron acceptors.
Bailey and Beauchamp (1973; 1973a) showed that the reduction of NO_3^- and NO_2^- occurred at redox potentials of 200 and 180 mV respectively under anaerobic conditions. The redox potential adapted itself to the cited values without any external adjustment and did not decrease until NO_3^- or NO_2^- was completely exhausted from the soil.
Although soil normally can be considered as a mixed culture, it is not clear whether the denitrification of soils is enzymatic or purely chemical. Van Cleemput and Patrick (1974) showed that the denitrification could occur in a water-logged soil even after a heavy γ-irradiation (2.5 Mrad). Thus, if denitrification is not a chemical process, it must be carried out by radiation-resistant enzyme systems of non-proliferating cells. However, the anaerobic denitrification in the sterile soil suspension was dependent on the redox potential with increasing nitrate/nitrite reduction for decreasing E_h values (Van Cleemput *et al.*, 1976).
Apart from being important to the denitrification processes, the redox potential also influences the reduction of iron and sulfate in soils (Engler and Patrick, 1973; Gotoh and Patrick, 1974).
The level of the redox potential depends on several factors in the soil. Meek and Grass (1975) investigated the effect of different factors on the redox potential in irrigated desert soils and found that a decrease in redox potential occurred as a result of (1) an increase in temperature (in the range 0–20 °C), (2) prolongation of the soil saturation time, and (3) higher amounts of substrate added to the soil. These findings all indicate that the changes in redox potential are due to microbial activity. Higher microbial activity is equivalent to increased demand for electron acceptors (e.g., O_2 or NO_3^- or NO_2^-), and therefore a higher microbial activity will lower the redox potential. It may therefore be concluded that the reactions that decrease the redox potential are of the same type for bioreactions in well controlled environments as for those under natural

conditions. Furthermore it has been postulated (Boichenko and Gryzhankova, 1974) that the redox potential of the atmosphere regulates CO_2 assimilation in the biosphere.

4.4 Redox Potentials in Other Systems

The redox potential has also been used in other systems as an indicator for microbial activity. Several authors have investigated the significance of the redox potential in sea water and in bottom sediments and have found that it was an important parameter when characterizing these media (Schmidt and Machan, 1975; Dechev *et al.*, 1974; Dechev *et al.*, 1974a).

In most of the studies cited above the redox potential has been used as an indicator for microbial activity. Therefore it seems obvious that the redox potential would be an important factor in killing microorganisms. In fact, the redox potential has been proposed as one of the parameters that might influence the disinfecting power of chlorinated water (Victorin *et al.*, 1972; Jentsch, 1973; Victorin, 1974).

4.5 Biochemical Fuel Cells

In Section 3.4 it was mentioned that normally no current is drained from the systems in which the redox potentials are measured. However, Allen (1966) proposed that a measurement of the current that can be drained from a microbial culture due to the potential differences that might be established could be a method that would yield more information about the metabolic processes.

From this idea it is only a small jump to the idea proposed by, e.g., Videla and Arvia (1975) that a microbial system could serve as the basis for a biochemical fuel cell. They investigated the strength of the current that could be obtained from a culture of *Saccharomyces cerevisiae* that was circulated through a bioelectrochemical cell and concluded that the system obeys simple phenomenologic laws; this finding could be useful for the development of bioelectrochemical devices for, e.g., electrochemical energy conversion. However, it must be emphasized that the energy that can be obtained in this way would be lower than the energy that could be obtained by combustion of the carbon source, and therefore this method of energy production would presumably only be advantageous in, e.g., equipment for waste-water treatment.

5. Regulation by Redox Potential

Although the significance of the redox potential for several biochemical reactions has been known for a long time, only a few researchers have investigated the possibilities of maintaining the redox potential at fixed values during prolonged periods.

As has been shown in the previous sections the concentration of dissolved oxygen is important for the level of the redox potential, although the real relation between DO and E_h is unknown. Lengyel and Nyiri (1965) described an automatic aeration control system using the redox potential (platinum electrode potential) to control the air inlet so that for a too-low E_h value the inlet rate of air was increased. This method is based on the fact that the k_La value increases for increasing inlet rate to a certain level (Miura,

1976). The method does not regulate the metabolism of the cells and is therefore a purely physical regulation.

Patrick *et al.* (1973) described a similar system for regulating the redox potential in soil suspensions. In the experiments of Lengyel and Nyiri (1965) and Patrick *et al.* (1973), the redox potential was in fact only regulated up-scale–namely, by adding more of the oxidant (i.e., O_2). Huang and Wu (1974) used the changes in the redox potential during growth of *Candida guillermondii* with n-paraffins as the carbon source to perform an iterative feeding of the substrate. By this method the additions regulate cellular metabolism. The redox potential decreased after each addition of the carbon source due to increased metabolism of the increased amount of cell mass and to the consequently increased oxygen demand, which for constant aeration implies a decrease in the concentration of dissolved oxygen.

From this it can be seen that this type of regulation is an operation that regulates metabolism and is in fact a down-scale regulation.

In our laboratory we have investigated the possibilities of establishing a continuous regulation of the redox potential through the automatic addition of glucose. A simple system for maintaining a constant redox potential by regulating the rate of glucose consumption is shown in principle in Fig. 1 (Kjaergaard and Joergensen, 1976).

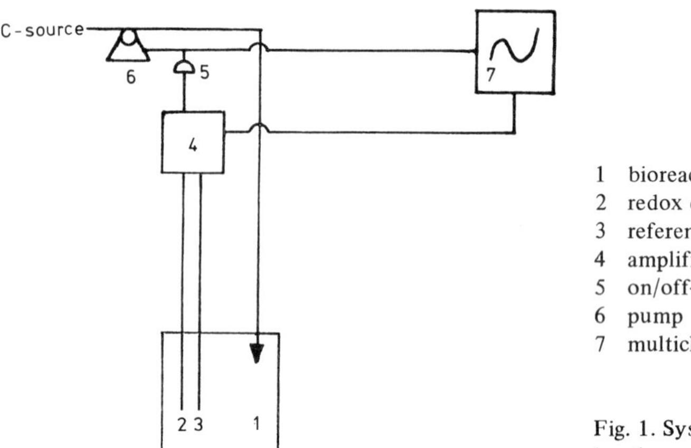

1 bioreactor
2 redox electrode
3 reference electrode
4 amplifier and amperemeter
5 on/off-relay
6 pump
7 multichannel recorder

Fig. 1. System in principle for regulating the redox potential in growth media

The apparatus, which is extremely simple, has been used to investigate the growth and metabolism of *B. licheniformis* at different constant values of E_h (Kjaergaard and Joergensen, 1977). It has been shown that it is possible to grow the bacteria at redox potentials of 240, 190, 140, and 90 mV for periods of up to 24 to 48 h. It was not possible to grow the bacteria at E_h = 40 mV, although the E_h during batch culture decreases to values as low as –30 mV. The reason for this may be that during a batch growth reductants can be accumulated in the medium in quantities that will not be produced when the concentration of the carbon source in the fermentor is low. The experiments con-

ducted at different values of E_h revealed the same connection between metabolism and redox potential as that obtained in chemostat experiments with *B. licheniformis* (Kjaergaard, 1976). Also the enzyme production (α-amylase) was found to depend on the redox potential.
However, the maintenance of a constant redox potential was not possible for periods longer than 24 to 48 h, presumably due to accumulation of metabolites and waste products that influenced the viscosity of the surface tension of the liquid, as a decrease of the k_La value was observed a few hours before the end of the period with constant E_h value. Because of this the specific oxygen uptake rate had to decrease and so the oxidative metabolism decreased. The reason for these changes is unknown.

6. Conclusions

Although the reactions and the components that determine the redox potential in microbial media are unkown, it is evident that the redox potential is a parameter that can provide valuable information about the status of microbial cultures, especially for cultures continuously supplied with oxygen. The dependence of the redox potential on the concentration of dissolved oxygen is significant, although the dependence can not be simply formulated using the Nernst equation.
Several experiments with submerse culture as well as with denitrification in soil indicate that the redox potential in the extracellular medium determines the effectivity of the processes taking place intracellularly. Therefore it can be recommended to consider the redox potential as a parameter that can be used for regulatory purposes. However, it must be emphasized that it is necessary to determine the characteristics with respect to redox potential of the system in which the redox potential is to be used as a regulating or a regulated parameter. It is not yet possible to put forward any general laws concerning the exact changes that take place in the redox potential during microbial growth, but the qualitative changes are well known.
Without doubt the research results in this area will bring the knowledge of redox potential up to a more satisfactory level during the next ten years.

7 . Acknowledgements

The author wishes to express his most grateful thanks to Dr. B. B. Joergensen for her valuable criticism and helpful discussions and also to the head of the department, professor O. B. Joergensen for making available the opportunity to prepare this manuscript.

8. References

Allen, M. J.: Symposium on bioelectrochemistry of Microorganisms. II. Electrochemical aspects of metabolism. Bact. Rev. **30**, 80–93 (1966).

Andreeva, E. A.: Physiological and biochemical rearrangements of the yeast *Candida utilis* as a function of the redox potential. Mikrobiologiya **43**, 780–785 (1974).

Bailey, L. D., Beauchamp, E. G.: Effects of temperature on NO_3^- and NO_2^- reduction, nitrogenous gas production, and redox potential in a saturated soil. Can. J. Soil Sci. **53**, 213–218 (1973).

Bailey, L. D., Beauchamp, E. G.: Effects of moisture, added NO_3^-, and macerated roots on NO_3^- transformation and redox potential in surface soils. Can. J. Soil Sci. **53**, 219–230 (1973a).

Balakireva, L. M., Kantere, V. M., Rabotnova, I. L.: The redox potential in microbiological media. Biotechnol. Bioeng. Symp. No. **4**, 769–780 (1974).

Blagova, N. V., Belozerova, A. V.: Redox potential as a differential sign of various types of *Shigella sonnei.* Zh. Mikrobiol. Epidemiol. Immunobiol. **47**, 59–62 (1970).

Boichenko, E. A., Gryzhankova, L. N.: Changes in iron compounds in the evolution of carbon dioxide assimilation. Zh. Evol. Biokkim. Fiziol. **10**, 119–122 (1974).

Cogdell, R. J., Jackson, J. B., Lofts, A. R.: The effect of redox potential on the coupling between rapid hydrogen-ion binding and electron transport in chromatophores from *Rhodopseudomonas spheroides.* Bioenergetics **4**, 211–227 (1973).

Dechev, G. D., Albert, H. O., Yordanov, S. D., Matveeva, E. G.: On the pH of sea water and sediments and its connection with the redox potential. Comp. Rend. Acad. Bulg. Sci. **27**, 559–562 (1974).

Dechev, G. D., Matveeva, E. G., Yordanov, S. D.: Determining the characteristics of sea water and bottom sediments as well as their interrelations by means of the redox potential. Comp. Rend. Acad. Bulg. Sci. **27**, 563–565 (1974a).

Einor, L. O.: Redox potential change in chloroplast suspension under illumination. Fiziol. Biokkim. Kul't. Rast. **4**, 630–634 (1973).

Engler, R. M., Patrick, Jr., W. H.: Sulfate reduction and sulfide oxidation in flooded soil as affected by chemical oxidants. Soc. Amer. Proc. **37**, 685–688 (1973).

Gemant, A.: Influence of the redox potential of the medium on the activity of polyphenol oxidase. Mol. Biol. Rep. **1**, 257–261 (1974).

Gillespie, L. J.: Reduction potentials of bacterial cultures and of water-logged soils. Soil Sci. **9**, 199–216 (1920).

Gotoh, S., Patrick, Jr., W. H.: Transformation of iron in a waterlogged soil as influenced by redox potential and pH. Soc. Amer. Proc. **38**, 66–71 (1974).

Hägg, G.: Kemisk reaktionslära, pp. 171–204. Stockholm: Almquist & Wiksell 1965.

Harrison, D. E. F.: Physiological effects of dissolved oxygen tension and redox potential on growing populations of micro-organisms. J. Appl. Chem. Biotechnol. **22**, 417–440 (1972).

Harrison, D. E. F.: Growth, oxygen, and respiration. CRC. Crit. Rev. Microbiol. **2**, 185–228 (1973).

Hays, R. C., Mandell, G. L.: pO_2, pH, and redox potential of experimental abscesses (38275) Proc. Soc. Exp. Biol. Med. **147**, 29–30 (1974).

Hill, R.: A note of some old and some possible new redox indicators. Bioenergetics **4**, 229–237 (1973).

Huang, S. Y., Wu, C. S.: Redox potential in yeast cultivation broth using n-paraffins as carbon source. J. Ferment. Technol. **52**, 818–827 (1974).

Ishizaki, A., Shibai, H., Hirose, Y.: Basic aspects of electrode potential change in submerged fermentation. Agr. Biol. Chem. **38**, 2399–2406 (1974).

Jacob, H.-E.: Das Redoxpotential in Bakterienkulturen. Z. Allg. Mikrobiol. **11**, 691–734 (1971).

Jacob, H.-E.: Reasons for the redox potential in microbial cultures. Biotechnol. Bioeng. Symp. No. **4**, 781–788 (1974).

Jentsch, F.: Redoxpotential und Keimtötung in gechlortem Meerwasser. Zbl. Bakt. Hyg. I. Abt. Orig. B. **157**, 304–312 (1973).

Joergensen, H.: Studies on the nature of the bromate effect. Thesis, pp. 105–113. Copenhagen: Munksgaard 1941.

Kantere, V. M.: Redox potential measurements in microbiological media and some applications. In: Millazzo, G., Jones, P. E., Rampazzo, L. (eds.): Exper. Suppl. **18**, 355–366 (1972).

Kjaergaard, L.: Influence of redox potential on the glucose catabolism of chemostat grown *Bacillus licheniformis.* Eur. J. Appl. Microbiol. **2**, 215–220 (1976).

Kjaergaard, L., Joergensen, B. B.: Maintenance of a constant redox potential during fermentation by automatical addition of glucose. In: Dellweg, H. (ed.): Abstracts of papers, p. 24. Fifth

International Fermentation Symposium. Verlag Versuchs- und Lehranstalt für Spiritusfabrikation und Fermentationstechnologie in Institut für Gärungsgewerbe und Biotechnologie zu Berlin (West) (1976).

Kjaergaard, L., Joergensen, B. B.: Regulation of redox potential by means of the glucose addition during batch fermentations of *Bacillus licheniformis.* In preparation (1977).

Linnet, N.: pH measurement in theory and practice, pp. 47–179. Copenhagen: Radiometer A/S 1970.

Lengyel, Z. L., Nyiri, L.: An automatic aeration control system for biosynthetic processes. Biotechnol. Bioeng. 7, 91–100 (1965).

Meek, B. D., Grass, L. B.: Redox potential in irrigated desert soils as an indicator of aeration status. Soil Sci. Soc. Amer. Proc. **39**, 870–875 (1975).

Miura, Y.: Transfer of oxygen and scale-up in submerged aerobic fermentation. In: Ghose, T. K., Fiechter, A., Blakebrough, N. (eds.): Advances in biochemical engineering. Vol. 4, pp. 3–40 (1976).

Onderdonk, A. B., Johnston, J., Mayleew, J. W., Gorbach, S. L.: Effect of dissolved oxygen and E_h on *Bacteroides fragilis* during continuous culture. Appl. Environ. Microbiol. **31**, 168–172 (1976).

Patrick, Jr., W. H., Williams, B. G., Moraghan, J. T.: A simple system for controlling redox potential and pH in soil suspensions. Soil Sci. Amer. Proc. **37**, 331–332 (1973).

Pirt, S. J.: The oxygen requirement of growing cultures of an *Aerobacter* species determined by means of the continuous culture technique. J. Gen. Microbiol. **16**, 59–75 (1957).

Schmidt, H.-E., Machan, R.: E_h-measurements in marine sediments under laboratory conditions. Cah. Biol. Mar. **16**, 733–741 (1975).

Schuldiner, S., Piersma, B. J., Warner, T. B.: Potential of a platinum electrode at low partial pressures of hydrogen and oxygen. II. An improved gas-tight system with a negligible oxygen leak. J. Electrochem. Soc. **113**, 573–577 (1966).

Shibai, H., Ishizaki, A., Kobayishi, K., Hirose, Y.: Simultaneous measurement of dissolved oxygen and oxidation-reduction potentials in the aerobic culture. Agr. Biol. Chem. **38**, 2407–2411 (1974).

Shikova, L., Panayotova, K., Deskova, G.: Cultural, morphological and biochemical investigations of species of the genus *Candida* and *Geotrichum candidum.* Ser. Sci. Med. Annu. Sci. Pap. **10**, 123–135 (1972).

Sukharevich, V. I., Yakovleva, E. P., Tsyganov, V. A., Shvezova, N. N.: The effect of aeration and redox potential of the medium on biosynthesis of levorin A and B. Mikrobiologiya **39**, 981–985 (1970).

Van Cleemput, O., Patrick, Jr., W. H.: Nitrate and nitrite reduction in flooded γ-irridiated soil under controlled pH and redox potential conditions. Soil Biol. Biochem. **6**, 85–88 (1974).

Van Cleemput, O., Patrick, Jr., W. H., McIlhenny, R. C.: Nitrite decomposition in flooded soil under different pH and redox potential conditions. Soil Sci. Amer. Proc. **40**, 55–60 (1976).

Victorin, K.: A field study of some swimming-pool waters with regard to bacterial, available chlorine, and redox potential. J. Hyg. Camb. **72**, 101–110 (1974).

Victorin, K., Hellström, K.-G., Rylander, R.: Redox potential measurements for determining the disinfecting power of chlorinated water. J. Hyg. Camb. **70**, 313–323 (1972).

Videla, H. A., Arvia, A. J.: The response of a bioelectrochemical cell with *Saccharomyces cerevisiae* metabolizing glucose under various fermentation conditions. Biotechnol. Bioeng. **17**, 1529–1543 (1975).

Wimpenny, J. W. T.: The effect of E_h on regulatory processes in facultative anaerobes. Biotechnol. Bioeng. **11**, 623–629 (1969).

Wimpenny, J. W. T.: Can culture redox potential be a useful indicator of oxygen metabolism by microorganisms? J. Appl. Chem. Biotechnol. **26**, 48–49 (1976).

Wimpenny, J. W. T., Necklen, D. K.: The redox environment and microbial physiology. I. The transition from anaerobiosis to aerobiosis in continuous cultures of facultative anaerobes. Biochim. Biophys. Acta **253**, 352–359 (1971).

Zobell, C. E.: Microbial activites and their interdependency with environmental conditions in submerged soils. Soil Sci. **119**, 1–2 (1975).

Zs.-Nagy, I., Ermini, M.: ATP production in the tissues of the bivalue *Mytilus galloprovincalis* (pelecypoda) under normal and anoxic conditions. Comp. Biochem. Physiol. **43B**, 593–600 (1972).

Zwarun, A. A.: Measurement of redox potential changes in anaerobic culture media caused by addition of blood. J. Lab. Clin. Med. **85**, 174–180 (1975).

Advances in Biochemical Engineering

Editors:
T. K. Ghose, A. Fiechter,
N. Blakebrough
Managing Editor:
A. Fiechter

Volume 1

70 figures. VII, 194 pages. 1971
ISBN 3-540-05400-6

Contents: The Nature of Fermentation Fluids. – Separation of Cells from Culture Media. – A Simplified Kinetic Approach to Cellulose-Cellulase System. – Production and Applications of Enzymes. – Overproduction of Microbial Metabolites and Enzymes Due to Alteration of Regulation. – The Production of Biomass from Hydrogen and Carbon Dioxide. – Liquid and Solid Hydrocarbons.

Volume 2

70 figures. V, 215 pages. 1972
ISBN 3-540-06017-0

Contents: Enzyme Engineering. – Application of Computers in Biochemical Engineering. – Mixed Microbial Populations. – Scale-Up of Biological Wastewater, Treatment Reactors. – Cellulose as a Novel Energy Source. – The Culture of Plant Cells

Volume 3

119 figures. VI, 290 pages. 1974
ISBN 3-540-06546-6

Contents: Seminar on Topics of Fermentation Microbiology. – Genetic Problems of the Biosynthesis of Tetracycline Antibiotics. – Some Aspects of Basic Genetic Research on Fungi and Their Practical Implications. – Microbial Oxidation of Methane and Methanol. – Modelling and Simulations in Biochemical Engineering. – Transient and Oscillatory States of Continuous Culture. – The Significance of Microbial Film in Fermenters. – Present State and Perspectives of Biochemical Engineering.

Volume 4: Engineering

87 figures. V, 172 pages. 1976
ISBN 3-540-07747-2

Contents: Transfer of Oxygen and Scale-Up in Submerged Aerobic Fermentation. – Microbial Flocs and Flocculation in Fermentation Process Engineering. – Analog/Hybrid Computation in Biochemical Engineering. – Preparation and Properties of Gel Entrapped Enzymes.

Volume 5: Microbial Products

31 figures, 27 tables. VII, 145 pages. 1977
ISBN 3-540-08074-0

Contents: Production of Cellulolytic Enzymes by Fungi. – An Evaluation of Enzymatic Hydrolysis of Cellulosic Materials. – Nucleic Acid Damage in Thermal Inactivation of Vegetative Microorganisms. – Cellular and Microbial Models in the Investigation of Mammalian Metabolism. – The Characterization of Mixing Fermenters.

Volume 6: New Substrates

28 figures. VII, 127 pages. 1977
ISBN 3-540-08363-4

Contents: The Role of Thiobacillus ferrooxidans in Hydrometallurgical Processes. – Cellulase Biosynthesis and Hydrolysis of Cellulosic Substances. – Metabolism of Methanol by Yeasts. – Control of Antibiotic Synthesis by Phosphate.

Springer-Verlag
Berlin
Heidelberg
New York

Advances in Polymer Science

Fortschritte der Hochpolymeren-Forschung

Springer-Verlag
Berlin
Heidelberg
New York

Volume 20: New Scientific Aspects

95 figures. IV, 227 pages. 1976
ISBN 3-540-07631-X
Contents: Syntheses, Conformation, and Reactions of Cyclic Peptides. – Properties of Liquid Crystals of Polypeptides – with Stress on the Electromagnetic Orientation. – ESR Studies on Polymer Radials Produced by Mechanical Destruction and Their Reactivity. – Catalytic Hydrolysis by Synthetic Polymers.

Volume 21: Mechanisms of Polyreactions – Polymer Characterization

68 figures. III, 151 pages. 1976
ISBN 3-540-07727-8
Contents: Poly (isobutylene-co-β-Pinene): A New Sulfur Vulcanizable, Ozone Resistant Elastomer by Cationic Isomerization Copolymerization. – Ring-Chain Equilibria and the Conformations of Polymer Chains. – Asymmetric Reactions of Synthetic Polypeptides. – Study of Polymers by Inverse Gas Chromatography.

Volume 22: Physical Chemistry

77 figures. III, 153 pages. 1977
ISBN 3-540-07942-4
Contents: Relaxation and Viscoelastic Properties of Heterogeneous Polymeric Compositions. – Electro-Optic Methods for Characterizing Macromolecules in Dilute Solution. – Ultrasonic Degradation of Polymers in Solution.

Volume 23: Reactivities

28 figures, 29 tables. III, 136 pages. 1977
ISBN 3-540-07943-2
Contents: Polymeric Reagents. Polymer Design, Scope, and Limitations. – Polymeric Drugs. – Cooperative Actions in the Nucleophile-Containing Polymers. – The Formation of Cyclic Oligomers in the Cationic Polymerization of Heterocycles.

Volume 24: Molecular Properties

130 figures, 33 tables. III, 244 pages. 1977
ISBN 3-540-08124-0
Contents: Polymer-Metal Complexes and Their Catalytic Activity. – Strain Energy Density Functions of Rubber Vulcanizates from Biaxial Extension. – ESCA Applied to Polymers. – Polymer Separation and Characterization by Thin-Layer Chromatography.

Volume 25: Polymer Chemistry

45 figures. 220 pages. 1977
ISBN 3-540-08389-8
Contents: Progress in the Chemistry of Polyconjugated Systems. – The Behaviour of Furan Derivatives in Polymerization Reactions. – Chemical Modifications of Fibre Forming Polymers and Copolymers of Acrylonitrile. – Synthetic Polyelectrolytes as Models of Nucleic Acids and Esterases.